Asia's Energy Trends and Developments

Case Studies in Cooperation, Competition and Possibilities from Central, Northeast and South Asia

volume 2

Asia's Energy Trends and Developments

Case Studies in Cooperation, Competition and Possibilities from Central, Northeast and South Asia

volume
2

Editors

Mark Hong
Asan Institute for Policy Studies, South Korea

Amy Lugg
Institute of Southeast Asian Studies, Singapore

World Scientific

NEW JERSEY • LONDON • SINGAPORE • BEIJING • SHANGHAI • HONG KONG • TAIPEI • CHENNAI

Published by

World Scientific Publishing Co. Pte. Ltd.

5 Toh Tuck Link, Singapore 596224

USA office: 27 Warren Street, Suite 401-402, Hackensack, NJ 07601

UK office: 57 Shelton Street, Covent Garden, London WC2H 9HE

Library of Congress Cataloging-in-Publication Data

Asia's energy trends and developments (in 2 volumes) / [edited by] Mark Hong, Asan Institute for Policy Studies, South Korea, Amy V.R. Lugg, Institute of Southeast Asian Studies, Singapore.

volumes cm

Includes index.

ISBN 978-9814425612 (Set)

ISBN 978-9814425575 (Vol. 1)

ISBN 978-9814425605 (Vol. 2)

1. Power resources--Asia. 2. Energy development--Asia. 3. Energy policy--Asia. I. Hong, Mark, editor of compilation. II. Lugg, Amy V. R., editor of compilation.

HD9502.A782A77 2013

333.79095--dc23

2013000742

British Library Cataloguing-in-Publication Data

A catalogue record for this book is available from the British Library.

In-house Editor: Lum Pui Yee

Typeset by Stallion Press

Email: enquiries@stallionpress.com

Printed in Singapore by B & Jo Enterprise Pte Ltd

FOREWORD

Volume 2 of this compilation is the result of a number of lectures that have been delivered at the ISEAS Energy Forum during 2009 and 2010. It takes a more regional approach, initially examining oil, gas and coal in Southeast Asia, then moving to East Asia and the need for cooperation in the face of competition for resources with Japanese and Russian energy relations. Next it examined the Japan-China relations in terms of energy policy. South Asia and specifically India's pipeline potential, which has links to the Caucasus and Central Asia are also explored, rounding off with the complexities of diplomacy among the Arab states relating to gas pipelines.

It is hoped that with this volume of the ISEAS Energy series of books, readers will have gained a comprehensive knowledge of all aspects of global and regional energy security. Energy issues are constantly moving and developing, being impacted by the latest political unrest in the Middle East or the latest energy technology development, oil prices rising, etc. It is necessary to update ourselves with an ever-changing energy story.

As with any publication that has multiple contributors, there are inevitable delays in collating the papers, therefore we wish to thank the patience and understanding of the contributors whose time they have contributed in order to help to enlarge public knowledge on vital energy issues.

Mark Hong & Amy Lugg
Editors
July 2012

CONTRIBUTORS

Ridwan D. Rusli conducts research in regulation theory and public economics at the Centre for Research in Economic Analysis (CREA), Faculty of Law, Economics and Finance, University of Luxembourg. He was formerly Research Fellow at Asia Competitiveness Institute, Lee Kuan Yew School of Public Policy, National University of Singapore. He is the founder of Exergy Advisory, an energy consulting firm, and advisor to Northstar Pacific, a private equity firm. Prior to that he was Managing Director, Head of Asia Natural Resources and Co-Head of Asia Energy at Credit Suisse Investment Banking, and energy specialist at UBS Investment Bank in Singapore, London, Zurich and Frankfurt. He was specialist advisor to state-owned and private companies as well as governments in Asia and Europe, and served on the boards of several energy, private equity and consultancy firms in Indonesia, China and Switzerland. His work included corporate strategy, mergers and acquisitions, project and capital market financings, state-owned company privatisations and industry restructurings. Mr. Rusli holds MSc degrees in Management and Chemical Engineering from MIT, and a Diplom Chemiker from Technical University of Berlin.
E-mail: ridwan.rusli@uni.lu

Christopher Len graduated from the University of Edinburgh with a MA Joint Honours degree in Philosophy and Politics. He then worked as Project Director in a Singapore market research firm doing consultancy work on social and market trends before leaving to pursue another Master's degree in the Peace and Conflict Research Department at Uppsala University in Sweden. He is presently Coordinator for the Energy and Cooperation project as well as the Conflict Management in Northeast Asia project for the Central Asia-Caucasus Institute and Silk Road Studies

Program. In addition, he is Assistant Editor for the *China and Eurasia Forum Quarterly* journal. Christopher Len is also the Central Asia editor for a new journal called the *Shingetsu Electronic Journal of Japanese-Islamic Relations* (SEJJIR), published by the Japan-based Shingetsu Institute, which focuses on Japan's relations with the Islamic world. He is also currently Visiting Associate at the Institute of Southeast Asian Studies (ISEAS) in Singapore. In 2000, he co-founded a community project in Kosovo for the local ethnic Roma minority focusing on post-conflict ethnic reconciliation between the Roma and the Albanian communities. On the side, in late 2006, together with four fellow Singaporeans, he completed a documentary called *I Love Malaya* (Asia Witness Production), about ex-Malayan communists presently exiled in Thailand.
E-mail: len.christopher@gmail.com

Svetlana Vassiliouk is a citizen of both Russia and the United States. She was born in Ukraine, grew up in Russia, moved to the US in 1994, and since 2002 has been living and working in Tokyo, Japan. She received her BA in Political Science summa cum laude in 1999 from Florida Atlantic University (Boca Raton, Florida); obtained her MA with distinction in International Relations and Economics, with a concentration in Japan Studies, in 2002 from Johns Hopkins University School of Advanced International Studies (SAIS) in Washington, DC; and was awarded a PhD in Political Science in 2006 from Hosei University in Tokyo. After graduating from Johns Hopkins SAIS in 2002, Dr. Vassiliouk moved to Tokyo. Upon her completion of the doctorate programme under the Monbukagakusho (Japanese Ministry of Education) Scholarship at Hosei University in 2006 (her dissertation was titled "Energy Politics in Japanese-Soviet/Russian Relations: From Cooperation Initiatives in the 1970s to Cautious Engagement in the 1990s"), she started her career as an adjunct professor of politics and international relations at Temple University, Japan Campus, and Hosei University, Graduate School of Politics. Dr Vassiliouk is currently Senior Assistant Professor at Meiji University, School of Global Japanese Studies, teaching courses in International Relations. While continuing to teach at Hosei University, she joined the Institute of Energy Economics, Japan, Prior to this, Dr Vassiliouk was an analyst in the Oil and Natural Gas Strategy Group at the Insitute of Energy Economics,

Japan and conducted research on Russian energy efficiency and conservation policies at the Asia Pacific Energy Research Centre (APERC). She delivered a number of conference papers and published several articles and a book chapter on a variety of topics related to Russian-Japanese energy cooperation, Russian-Ukrainian energy relations, Russian foreign policy, US-Japanese relations and others.
E-mail: vassiliouk@aperc.ieej.or.jp

Yuji Morita is Senior Research Fellow and Director at the Energy Data and Modelling Center (EDMC), Institute of Energy Economics, Japan. Mr. Morita joined the Institute of Energy Economics, Japan, (IEEJ) in 1998 as Senior Economist of the Team for Strategic Research Projects. In 2005, he was appointed Senior Research Fellow and Director at the EDMC. His responsibilities cover project management for various kinds of research, analysis and forecast projects on energy demand, supply, energy policies and energy technologies. Prior to joining IEEJ, he started his career at Japan Energy Corporation, where he held a variety of engineering and commercial posts. From 1995 he served as Senior Manager of the Research and Planning Department, responsible for the company's upstream businesses in China, Russia and the Middle East region. He received his BSc from the University of Kyoto, majoring in Organic Chemistry.
E-mail: ymorita@edmc.ieej.or.jp

Marie Lall, is a South Asia expert (India, Pakistan and Burma/ Myanmar) specialising in political issues (with regard to the economy, geopolitics of energy, foreign policy formulation, citizenship and diaspora politics) and education (with specific regard to education policy, gender, ethnicity and social exclusion issues, the formation of national identity and its close links with citizenship). She has 18 years of experience in the region, conducting extensive fieldwork and having lived both in India and Pakistan. She has written widely on these topics and is the author/editor of four books, with two more forthcoming. Dr. Lall has advised the German Ministry for Development in Berlin (on Pakistan); the Head of the German Foreign Relations Committee (on Pakistan and Myanmar); the Japanese Ministry of Foreign Affairs (MOFA) (on the Myanmar elections); Chinese, Canadian and Norwegian ministerial/embassy staff (on India,

Pakistan and Myanmar); and is recognised in the UK as a country expert by the Asylum and Immigration Court. She is an education advisor to The Citizens Foundation (TCF), Pakistan, and education specialist at Myanmar Egress. She appears regularly on BBC News 24/BBC World, Sky TV, Reuters TV, Al Jazeera TV and BBC Radio and gives interviews on politics and international relations in South Asia. She has been cited in international press on India, Pakistan and Myanmar. Some significant appearances were Al Jazeera's *Inside Story on the Ayodhya Verdict* (2010), BBC World Service's *The World Today* (2009) and the *Riz Khan Show* after the Mumbai attacks (2008). Her non-academic speaking engagements include keynote addresses at the Suntory Foundation, Osaka; the Heinrich Boell Stiftung, Berlin; the House of Lords; and the European Commission. She lectures at the Institute of Education, University of London. She is also an associate fellow on the Asia Programme at Chatham House and an honorary fellow at the Institute of South Asian Studies, National University of Singapore. She has had a number of short fellowships at world-renowned universities in Australia, Germany, India and Pakistan. She received her PhD from the London School of Economics in 1999.
E-mail: M.Lall@ioe.ac.uk

Hooman Peimani, is Head of the Energy Security Division at the Energy Studies Institute (ESI), National University of Singapore. Drawing on his years of work experience with academic, non-academic, private, public, national and international institutions in North America, Europe and West Asia, including UN agencies, he specialises in energy security, particularly relating to South and West Asia. Having over 20 years of research experience, his extensive publications include books, chapters in books, journal/newspaper articles, government/UN documents/reports and book reviews. He has also contributed as an expert to the publications of many news agencies (e.g. Reuters and United Press International) and/or been quoted by them. Since 1997, in the capacity of an expert in energy and security, he has made regular contributions to the programmes and publications of Radio Free Europe/Radio Liberty and the programmes of Radio France Internationale and Deutsche Welle.
E-mail: esihp@nus.edu.sg

Zhanibek Saurbek is a researcher from Kazakhstan and obtained his doctoral degree at the Centre for Energy, Petroleum and Mineral Law and Policy (CEPMLP), University of Dundee, Scotland, UK, in 2009. He holds a Bachelor's degree from the Kazakh State Juridical University (with distinction; 1998) and obtained a Candidate of Legal Science (Kandidat Nauk) with specialisation in the theory and history of law and the state in 2001. His sphere of interests includes the oil and gas pipeline industry, cross-border issues, and legal and commercial questions about the transportation infrastructure. Currently, he works in the oil sector of Kazakhstan in the KazMunaiGas system. Prior to joining this team, he spent four years working in the oil and gas sector in Kazakhstan, in particular in the pipeline projects Kenkiyak-Atyrau and Atasu-Alashankou (the Kazakhstan-China Pipeline). Dr. Saurbek prepared an issue for the *Oil, Gas and Energy Law* journal on Eurasian energy and has published several papers in various journals. He speaks Kazakh, Russian and English.
E-mail: zsaurbek@gmail.com

Mary E. Stonaker, formerly with the Middle East Institute in Singapore, focused on energy security, specifically that between the Middle East, South Asia and Latin America. Recognising the rising global importance of natural gas, Mary has begun making an acute focus towards analysing the political and economic dimensions of natural gas energy security. She studied International Relations with a specialisation in Foreign Policy and Security Studies at Boston University. Interested in linguistics, Mary has an ear for Arabic and Spanish. Mary has written for the Middle East Institute and given a seminar at the Institute of South Asian Studies in Singapore as well as contributing to *The Straits Times* and various international media outlets. Capitalising on the wealth of intellectual exchange in Singapore, Mary absorbs and shares on policy with unbiased openness.
E-mail: marystonaker@nus.edu.sg

Mark Hong was born in Singapore and educated at Raffles Institution, Singapore. After completing his secondary school education, Mr. Hong was awarded a President's Scholarship in 1965 and a Humanities Scholarship and Best Entrant scholarship at Singapore University. He obtained a BA in Economics from Cambridge University in 1969 and an MS in International Relations from Georgetown University in 1982 on a Fulbright Scholarship.

Mr. Hong joined the Ministry of Foreign Affairs in 1969. He served at the Singapore Embassy in Phnom Penh as Charge d'Affaires (1974–1975), at the Singapore Commission in Hong Kong as First Secretary (1975–1976), at the Singapore Embassy in Paris as Counsellor (1982–1986) and at the Singapore Permanent Mission to the United Nations in New York as Deputy Permanent Representative (1988–1994). At the Ministry of Foreign Affairs headquarters, he has served in various senior capacities as the director of several departments. His last foreign posting was as Singapore's Ambassador to Russia and Ukraine from November 1995 to March 2002. From May 2002 to January 2004, he was attached to the Institute of Defence and Strategic Studies, Nanyang Technological University, Singapore, as a Visiting Senior Fellow.

He is currently a Vice-Chairman of the International Committee of the Singapore Business Federation, and was a Visiting Research Fellow at ISEAS from February 2004 to October 2011. After his retirement in October 2011, he was appointed Visiting Senior Lecturer at James Cook University on its Australia and Singapore campuses, and also at Capilano University in Vancouver, and as Senior Visiting Fellow at Asan Institute in Seoul.

He has delivered over 300 papers and lectures to various international seminars and conferences, and attended many UN General Assemblies, ASEAN conferences and other regional meetings. He has edited 10 books for ISEAS, five on energy issues, three forthcoming in 2012, three on ASEAN-Russia relations, and one each on Southeast Asia and Cambodia. He has also contributed many chapters to various books, given lectures to groups such as the Young PAP, schools, and visiting delegations. He is writing a novel and several books of essays.
E-mail: markhong@gmail.com

Amy V. R. Lugg is responsible for public information and communications for the ASEAN Studies Centre at the Institute of Southeast Asian Studies (ISEAS). Before joining the Centre, she was Associate Editor with a leading provider of energy and metals information. Prior to that,

she was Executive Search Director at a boutique executive search company in Singapore with extensive networks in the banking and finance, academia, hospitality, retail and energy sectors. Amy has also been a Visiting Associate at ISEAS since June 2009, with the ISEAS Energy Studies Programme. She is co-editor of the ISEAS Energy Series publications with Mark Hong. Amy holds a Master's degree in International Relations from Curtin University, Australia, and her research interests include energy security, human security and transnational crime.
E-mail: amy@iseas.edu.sg

CONTENTS

CHAPTER 1

THE LOGIC OF ENERGY POLICY: THE CASE OF UPSTREAM OIL, GAS, COAL AND DOWNSTREAM OIL SECTORS IN SOUTHEAST ASIA

Ridwan D. Rusli

1.1 INTRODUCTION

Energy inputs such as electricity and fuels are essential to facilitate industrial, commercial and household activities. The characteristics of electricity and fuel demand are determined by the objectives of the main users: households need affordable, steady and safe availability, while commerce and industry demand an economically competitive, secure and sustainable supply of electricity and fuel. Thus the price and availability of energy are important determinants of economic competitiveness (through lower inputs, production and transport costs), energy supply security (to ensure stable industrial, commerce and household activities) and sustainable development (through sustainable management of natural resources, energy efficiency, environmental and climate control measures).[1]

[1]The Ministry of Trade and Industry of Singapore, in its recent *National Energy Policy Report*, listed economics, security and environment as the three objectives of energy

Among primary energy feedstocks, fossil fuels make up the largest components of total primary energy supply (TPES). The International Energy Agency (IEA) forecasts that the share of oil, natural gas and coal in world TPES in 2030 will range from 67.1% to 80.5%.[2] In the year 2008, the share of oil, natural gas and coal in Southeast Asian TPES ranged from 91.1% to 100%, compared to 85.7% in Japan, 92.6% in China and 93.2% in India.[3] While the share of fossil fuels in TPES is expected to be higher in developing countries, Southeast Asian economies' near total reliance on oil, natural gas and coal is much higher than the average of 83.6% in the more developed European Union, 85.4% in OECD[4] countries and 87.0% in North America. Interestingly, the share of fossil fuels in Latin America of 72.9% is considerably lower, which could be partially a result of its significant use of hydroelectricity.

Given the critical importance of oil, natural gas and coal to the industrial and economic development of Southeast Asian countries, fossil fuels play a critical role in regional- and national-level energy policy. Oil, natural gas and coal endowments have a critical impact on the economic competitiveness, socio-political security and stability, as well as sustainable development in the region. It is therefore disconcerting that the region, after decades of having Indonesia and Malaysia among the world's important crude oil exporters, has seen Indonesia become a net importer of oil, despite the country's resource potential. Although Singapore re-exports a significant amount of the fuel products from its world-class refineries, the region

---

policy. We elaborate on these three objectives in Figure 1 of Section 1.2 Objectives of Energy Policy. See Singapore Ministry of Trade and Industry (MTI), *National Energy Policy Report* (Singapore: MTI, 2008), p. 23.

[2] The IEA defines two scenarios: a Reference Scenario based on current policies and a 450 Policy Scenario based on a plausible post-2012 climate-policy framework to stabilise the concentration of global greenhouse gases at 450 parts per million (ppm) carbon dioxide (CO_2) equivalent. See IEA, *Key World Energy Statistics 2009* (Paris Cedex: IEA, 2009), p. 47.

[3] Southeast Asian countries listed in BP's TPES table include Indonesia, Malaysia, Thailand, the Philippines and Singapore. Average share of TPES of oil, gas and coal for the entire Asia Pacific was 91.7% in 2008. See BP, *Statistical Review of World Energy June 2009* (London: BP, 2009), p. 41.

[4] Organisation for Economic Co-operation and Development.

will import larger volumes of crude oil in the future.[5] The region's hope thus rests with future crude oil production growth from newly developed or yet-to-be-discovered fields in Vietnam, Malaysia and Indonesia.

The situation looks comparatively less worrisome in natural gas and coal. In natural gas, Indonesia and Malaysia continue to be among the world's largest exporters of liquid natural gas (LNG). The largest LNG consumers are Japan, South Korea and Taiwan. Nevertheless, while new future demand could originate from China, India and the Pacific West Coast, competition from Australia, the Middle East, e.g. Qatar, enhanced by the development of a global LNG spot market, as well as domestic gas resources in India and China, would intensify competition in the regional LNG market.[6] Thailand has successfully built an extensive infrastructure for the domestic production and transport of natural gas, its most abundant domestic resource.[7] Myanmar, historically a limited exporter of gas to Thailand, has discovered several large gas fields that can be developed in the course of the next decade. At the same time extensive exploration activity and several discoveries have been observed in Vietnam and, more recently, Cambodia.[8] In coal, though Indonesia was the world's largest exporter of thermal coal for power generation in 2008, Australia's production capacity is continuously being

[5] See Ong Eng Tong, "The Singapore Oil Situation", in ISEAS, *Energy Perspectives on Singapore and the Region* (Singapore: ISEAS, 2007); Cheng Hong Kok, "Singapore Petroleum Company: Adding Value to the Singapore Oil Industry", in ISEAS, *Energy Perspectives on Singapore and the Region*; Yoshikuni Yamakawa, "Outlook of Products Supply & Demand in Asia by Country and by Product", presented at Northeast Asia Petroleum Forum 2009 (Tokyo: Institute for Energy Economics, Japan (IEEJ), 2009).

[6] For reviews of regional and world gas and LNG markets, see IEA, *Natural Gas Market Review 2008* (Paris Cedex: OECD/IEA, 2008); and Allison Ball, Karen Schneider, Lindsay Fairhead and Christopher Short, *The Asia Pacific LNG Market, Issues and Outlook* (Canberra: Australian Bureau of Agricultural and Resource Economics, 2004).

[7] See National Energy Policy Council (NEPC), *Thailand's Energy Policy and Development Plan under the Administration of Prime Minister General Surayud Chulanont* (Bangkok: NEPC, 2006); and PTT Plc, *Annual Report 2008* (Bangkok: PTT Plc, 2009).

[8] See Pou Sothirak, "An Overview of the Cambodian Energy Sector", in *Energy Essays on the Asia-Pacific Region*, co-edited by Amy V. R. Lugg and Mark Hong (Singapore: ISEAS, 2009), pp. 57–96.

increased.[9] In both natural gas and coal Southeast Asian exporters cannot afford to become complacent, since competition from Australia and the Middle East will intensify in the future.

Energy policy must facilitate environmentally sustainable development and growth of electricity and fuel supply infrastructure. How governments can best develop and implement energy strategies and policies to achieve the three goals of energy policy, namely economic efficiency and competitiveness, energy supply security and environmental sustainability, is a function of each country's demographics (total and per capita income, education, lifestyle), geography (natural resource endowment, location of demand versus supply centres, country size and geographic spread, environmental factors), socio-political (sector development and history, social and political institutions), as well as industrial and economic structure (manufacturing, agricultural versus service sectors, entry and competition policy, trade, investment and fiscal policy). The goal of supply-side energy policy in Southeast Asia should be to reduce import dependence and increase fuel sourcing flexibility in crude oil, to maximise cost efficiency and competitiveness across the oil, gas and coal value chains, i.e. exploration, production and processing, to develop an efficient transport and distribution infrastructure, as well as to diversify electricity and fuel production using non-fossil fuel and alternative energy sources to help preserve the environment. On the demand side, efficiency improvement measures across commercial, industrial and household activities, as well as energy conservation programmes, will require extensive investments in education and infrastructure.

This chapter describes a logical, normative framework for energy policy making in selected Southeast Asian countries, with emphasis on the fossil fuel–based oil, natural gas and coal industries. It focuses on the interaction between two of the three main objectives of energy policy, economics and security. This normative framework can be extended to other alternative energy sources and fuel types, as well as to target more specifically the interdependence between economic competitiveness, energy security and politics, as well as environmental sustainability. The emphasis is on the

---

[9] Energy Information Administration (EIA), *International Energy Outlook 2009* (Washington, DC: EIA, 2009), p. 55.

growing Southeast Asian economies that, not unlike China and India, are becoming increasingly dependent on imported energy feedstock, particularly crude oil. It complements the already rich literature on Southeast Asia's increasing energy import dependence, vulnerability to high oil price shocks, energy security and supplier concentration risk, as well as interaction with other large energy consumers across Asia.[10–12]

---

[10] For summaries of Southeast Asian energy sector challenges and energy security, refer to Dr. Andrew T. H. Tan, "The ASEAN Countries' Interest in Asian Energy Security", in Lugg and Hong, *Energy Essays on the Asia-Pacific Region*, pp. 1–20; Vince S. Pérez, "Who Wins the Asian Scramble for Oil?", in ISEAS, *Energy Perspectives on Singapore and the Region*, pp. 249–265; Steve Puckett, "The Outlook for Gas in Southeast Asia", in *Energy Perspectives on Singapore and the Region*, pp. 317–336; Kang Wu and Caleb R. O'Kray, "The Implications and Impacts of China's Oil Demand on the Asia Pacific", in ISEAS, *Energy Perspectives on Singapore and the Region*, pp. 142–155; Christopher Len, "Energy Security Cooperation in Asia: An ASEAN-SCO Energy Partnership?", in *Energy Perspectives on Singapore and the Region*, pp. 156–175; Mark Hong, "New Partnerships in Energy Security in Asia: Between India, ASEAN and Singapore", in Lugg and Hong, *Energy Essays on the Asia-Pacific Region*, pp. 113–131.

[11] For a comprehensive overview of the state of world energy, see Vaclav Smil, *Energy at the Crossroads: Global Perspectives and Uncertainties* (Cambridge: The MIT Press, 2003); for general discussions on energy security indicators, see Nicolas Lefèvre, "Measuring the Energy Security Implications of Fossil Fuel Resource Concentration", *Energy Policy*, 38 (2010), pp. 1635–1644; and Andreas Löschel, Ulf Moslener and Dirk T. G. Rübbelke, "Energy Security: Concepts and Indicators", *Energy Policy*, 38 (2010), pp. 1607–1608.

[12] For overviews of sector challenges and energy security in China, India and Japan, see Henry Leong, "China's Energy Security: Geopolitics versus Interdependence", in ISEAS, *Energy Perspectives on Singapore and the Region*, pp. 176–196; Dr. Wenran Jiang, "China's Global Quest for Energy Security", in Lugg and Hong, *Energy Essays on the Asia-Pacific Region*, pp. 132–164; Michael Richardson, "Energy & Geopolitics in the South China Sea", in Lugg and Hong, *Energy Essays on the Asia-Pacific Region*, pp. 165–191; Ligia Noronha, "India's Energy Situation: The Need to Secure Energy Resources in an Increasingly Competitive Environment", in ISEAS, *Energy Perspectives on Singapore and the Region*, pp. 132–141; Rajiv Sikri, "India's Energy Challenges", in Lugg and Hong, *Energy Essays on the Asia-Pacific Region*, pp. 97–112; Hisane Masaki, "Japan's New Energy Strategy", in ISEAS, *Energy Perspectives on Singapore and the Region*, pp. 228–248; Yuji Morita, "Japan's Energy Supply-Demand Situation and Energy Conservation Policy and Challenges for Japan", in Lugg and Hong, *Energy Essays on the Asia-Pacific Region*, pp. 205–236.

The section following this introduction describes the main objectives of energy policy. The next section discusses the demographic, geographic, socio-political and economic determinants of energy policy, according to which countries can be categorised into several groups. The subsequent section summarises the key instruments of energy policy and identifies a possible normative grid of energy strategies and policies consistent with the characteristics of each country grouping. The normative policy grid derived is compared in the next section with the actual energy policies observed in Indonesia, Thailand and Malaysia. This comparative analysis of normative versus observed policy instruments includes industry regulation, trade, investment and fiscal policy, as well as the institutional dynamics and implementation track record across the selected Southeast Asian countries. The final section summarises and concludes. Lastly, the endnotes provide commentaries, guidance to references and other relevant works.

1.2 OBJECTIVES OF ENERGY POLICY

As depicted in Figure 1, the three main goals of energy policy are interdependent and must be addressed simultaneously. Economic competitiveness, supply security and flexibility are driven by several common factors. The availability of energy feedstock such as crude oil, natural gas and coal at competitive prices is required to ensure the competitiveness of industrial, agricultural and commercial activities, and to help increase household purchasing power and living standards. Supply security and flexibility can be addressed through fuel diversification strategies, while efficient transport and logistics infrastructures are needed to bring energy feedstock to the domestic fuel and electricity generating companies, as well as manufacturing goods and agricultural products to domestic consumers and export markets. Furthermore, efficient, environmentally sustainable and reliable technologies, facilities and operations are equally important to convert energy feedstock and production inputs into marketable products that can be enjoyed by the consumers. To ensure appropriate incentives and rewards for industrial, commercial and household

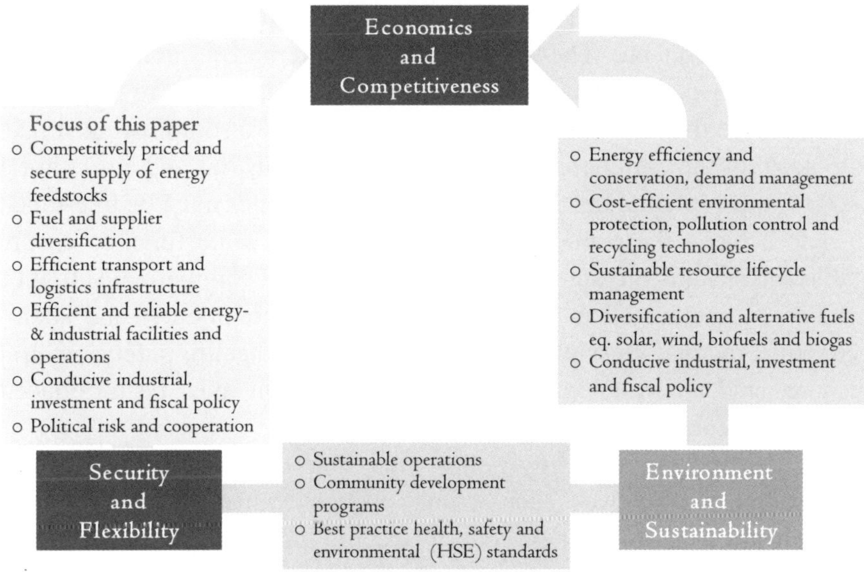

Figure 1 Three Objectives of Energy Policy

Source: Author's analysis based on the policy objectives laid out in Singapore Ministry of Trade and Industry (MTI), National Energy Policy Report (Singapore: MTI, 2008), p. 23.

investments of capital and effort into the realisation of these energy-related projects, supportive industrial, investment and fiscal policies, together with a functioning financial sector, also contribute to the overall economic competitiveness and socio-economic development of a country.

Economic competitiveness and environmentally optimal management of natural resources and industrial pollution are crucial to ensure long-term and sustainable industrialisation, income growth and socio-economic development. Energy efficiency and demand management must be encouraged and promoted, not only to reduce cost and improve competitiveness, but also to ensure long-term sustainable growth in productive output and income levels.

Measures such as optimal and cost-effective energy conservation, energy efficiency and pollution control technologies and work processes, as well as optimal natural resource extraction and lifecycle management[13] must be undertaken. The introduction of alternative, non-fossil fuel energy inputs such as solar, wind, biofuels and biogas need to be encouraged.[14] Since the implementation of these alternative solutions must take into account the commercial objectives of the industrial and agricultural sectors, appropriate industrial, trade and fiscal policies must be fine-tuned to help achieve these multiple objectives. At the same time, to ensure stable and sustainable output growth, industrial and commercial activities need to be conducted in consideration of local community and socio-economic factors. Best practices in the areas of health, safety, natural resource and environmental pollution management must be introduced and implemented.

In summary, while energy policy must take into account these three interdependent objectives, each country and government must tailor their policy instruments to fit their country-specific historical, institutional and natural environments. A country's or region's determinants of energy policy can be examined along four dimensions, which are characterised by distinct demographic, geographic, socio-political and economic-industrial factors.

1.3 DETERMINANTS OF ENERGY POLICY

The choice of optimal energy policy instruments depends on a combination of supply- and demand-side factors. On the supply side, geography and resource endowment are key, the latter encompassing oil, gas and coal

---

[13] For conceptual overviews of the sustainable management of mining and extractive industries, see Mining, Minerals and Sustainable Development North America (MMSD), *Toward Change: The Work and Results of MMSD-North America* (Winnipeg: International Institute for Sustainable Development (IISD), 2002); and Eleodoro Mayorga Alba, "Extractive Industries Value Chain: A Comprehensive Integrated Approach to Developing Extractive Industries", *Extractive Industries for Development Series*, no. 3 (Washington, DC: The World Bank, 2009).

[14] For a broad overview of non-fossil energies see Smil, *Energy at the Crossroads: Global Perspectives and Uncertainties*, pp. 239–316; while biofuels offer promising fuel diversification potential, especially for agriculturally endowed Southeast Asia, the debate regarding its impact on future arable land for food production is ongoing. See for example David Pimentel *et al.*, "Food versus Biofuels: Environmental and Economic Costs", *Human Ecology*, 37 (2009), pp. 1–12.

reserves.[15] On the demand side, demographic, geographic, economic and socio-political factors determine a country's primary energy mix. While demographic factors include age, income distribution, industrial and household behaviour, each country's geography influences the spatial distribution of natural resources, population and economic activity, as well as the need for factor mobility and transportation of semi-finished and finished manufacturing goods and agricultural products.

Socio-political factors include for example a government's and society's bias towards consumer welfare (versus firm or industry profits), which is reflected in the choice of industrial policies,[16] and the importance of environmental protection in that country. Together with an economy's unique pattern of agricultural, industrial, investment, consumption and financial sector activities, the demographic, geographic and socio-political characteristics determine the demand for transportation fuel and electricity.

All these factors are critical to the prioritisation of energy policy objectives and formulation of a normative energy sector strategy. In particular, most countries face a certain imbalance between resource endowment and primary energy consumption. Some countries end up becoming net importers, others net exporters of fuel and energy.[17] In terms of prioritisation of objectives, fuel- and energy-importing countries are typically more concerned about energy supply security, fuel diversification and economic competitiveness. Generally they will be focused on energy conservation, improving energy efficiency and industrial competitiveness as a buffer against unpredictable world energy supply and prices. On the other hand, fuel- and energy-exporting countries concentrate their efforts on maximising resource production, improving international competitiveness and promoting world trade. Lastly, some countries, especially in both fuel- and energy-importing groups, have

---

[15] This chapter focuses on key fossil fuel inputs for transportation fuel, electricity generation and petrochemicals, as relevant.

[16] The formulation and implementation of industrial and energy policy is of course significantly influenced by the political actors and decision makers (see discussion in section 1.3.7 Regional Summary).

[17] Oil and coal are generally more easily tradable and transportable, natural gas is mainly traded through long-term pipeline (usually shorter distances) or LNG contracts (short and long distances). While active markets for pipeline gas exist in the more inter-connected Western Europe and North America, the global spot market for LNG is only gradually developing.

successfully shifted their focus towards environmental protection and alternative energy technologies and power generation.[18]

Normatively, each country's government will direct investment, industrial development, technology and skill accumulation in accordance with its energy priorities and strategies. However, to ensure the effective implementation of a country's energy strategy, three more factors are of utmost importance. First, a set of detailed implementation rules and regulations are required to provide guidelines to the country's energy sector and related institutions.[19] Next, the investment and funding environment is crucial. In many cases, energy sector investments necessitate the concurrent development of efficient funding and capital markets, and under certain conditions, targeted fiscal subsidies. Third, as will be discussed later, the formulation of an energy strategy and its associated implementing rules and regulations needs to take into account, and is largely determined by, the country's political and institutional foundations.

1.3.1 Geography and Resource Endowment

The proven crude oil reserves in Southeast Asia are concentrated in Malaysia, Vietnam, Indonesia and to a lesser extent in Brunei and Thailand (Figure 2). Indonesia and Malaysia have historically been net exporters of crude oil, but due to declining domestic crude oil production and growing domestic demand for fuel products, Indonesia has become a net crude and fuel product importer while Malaysia, which faces a growing deficit in refined products, is still able to balance its net fuel imports with net crude exports, at least for another few years (Figure 3). Vietnam's growing role as the region's fourth-largest oil producer is a result of the

[18] Examples include countries such as Japan, and in Southeast Asia, to a lesser extent, the Philippines.

[19] This chapter interchangeably uses the term "institution" to represent both institutions and organisations. North defines "institutions" as "the rules of the game in a society or humanly devised constraints that shape human interaction", and "organisations" as "groups of individuals bound by some common purpose to achieve objectives". Thus organisations could include political, economic, social and education bodies. See Douglas C. North, *Institutions, Institutional Change and Economic Performance* (Cambridge: Cambridge University Press, 1990).

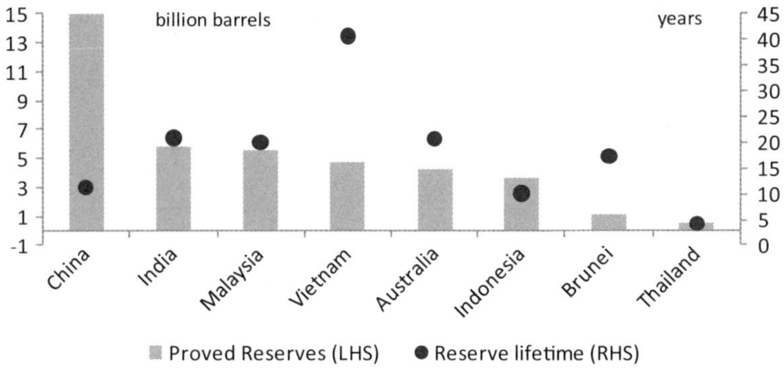

Figure 2 Asia Oil Reserves and Lifetime 2008

Source: *BP Statistical Review of World Energy*, June 2009.

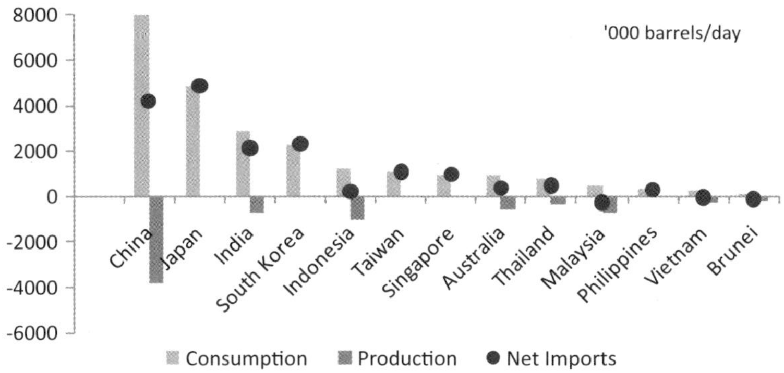

Figure 3 Asia Oil Consumption, Production and Net Imports 2008

Source: *BP Statistical Review of World Energy*, June 2009; EIA, *International Energy Outlook*, 2009.

significant increase in oil and gas exploration activity in recent years, with potential for new oil and gas discoveries.[20]

The region will need to import larger volumes of crude oil from the Middle East in the future. Its hope thus rests with future crude oil production growth from newly developed or yet-to-be-discovered fields in Vietnam, Malaysia and Indonesia. Singapore, the most industrialised

[20] Energy Information Administration (EIA), *Vietnam: Country Analysis Briefs* (Washington, DC: EIA, 2007).

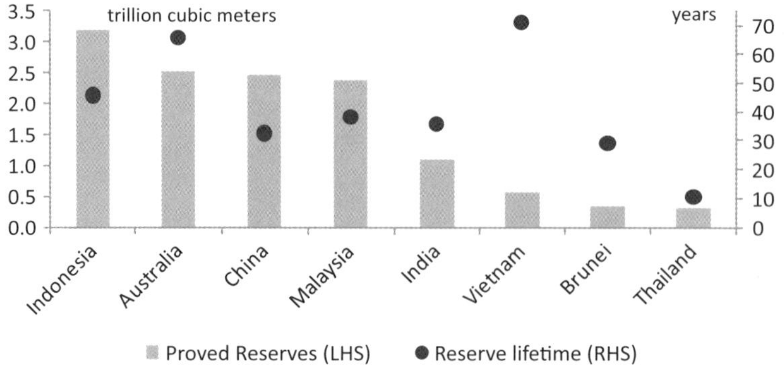

Figure 4 Asia Natural Gas Reserves and Lifetime 2008

Source: *BP Statistical Review of World Energy*, June 2009.

country in Southeast Asia, imports all of its crude oil and natural gas requirements, and re-exports a significant amount of the fuel products from its world-class refineries. As will be elaborated later, the city state has successfully developed the third-largest oil refining centre in the world.

The situation looks comparatively less worrisome in natural gas and coal. The region's existing proven natural gas reserves are located in Indonesia, Malaysia, Vietnam and Brunei (Figure 4). Indonesia and Malaysia are among the world's top three exporters of liquid natural gas (LNG), with growing domestic gas demand, while Brunei exports LNG as well (Figure 5). The largest LNG consumers are currently Japan, South Korea and Taiwan. While additional future demand could arise from China, India and the Pacific West Coast, competition from Australia, the Middle East, e.g. Qatar, enhanced by the development of a global LNG spot market, as well as domestic gas resources in India and China, would intensify competition in the regional LNG market.[21]

Myanmar, historically a limited exporter of gas to Thailand, has seen several major gas discoveries, although commercialisation and start-up will take some time. Its gas discoveries will only be categorised as proven once clear off-take contracts and development plans are in place. Vietnam

[21] IEA, *Natural Gas Market Review 2008*; and Ball *et al.*, *The Asia Pacific LNG Market, Issues and Outlook*.

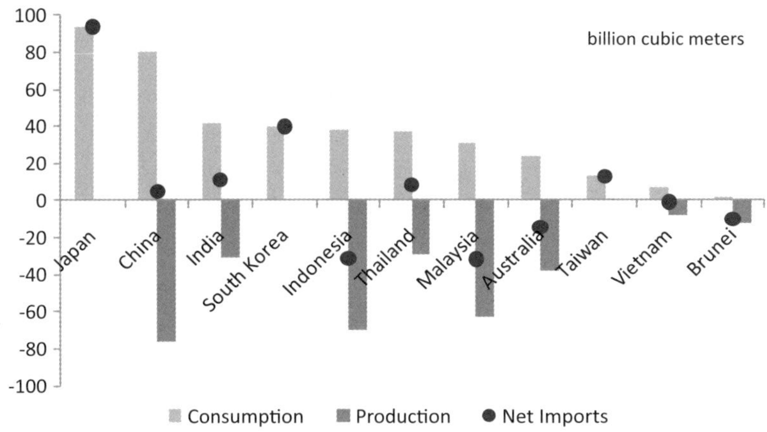

Figure 5 Asia Natural Gas Consumption, Production and Net Imports 2008

Source: *BP Statistical Review of World Energy*, June 2009; EIA, *International Energy Outlook*, 2009.

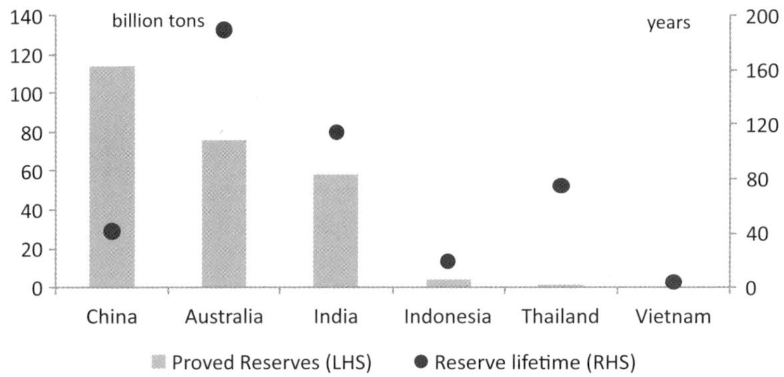

Figure 6 Asia Coal Reserves and Lifetime 2008

Source: *BP Statistical Review of World Energy*, June 2009.

is also expected to experience rapid growth in gas production, starting from its currently low base (Figure 6). Significant exploration activity is being conducted by several international oil majors and exploration and production (E&P) independents. Thailand is already a significant domestic consumer and producer of natural gas, and has successfully built an extensive infrastructure for the domestic production and transport of its

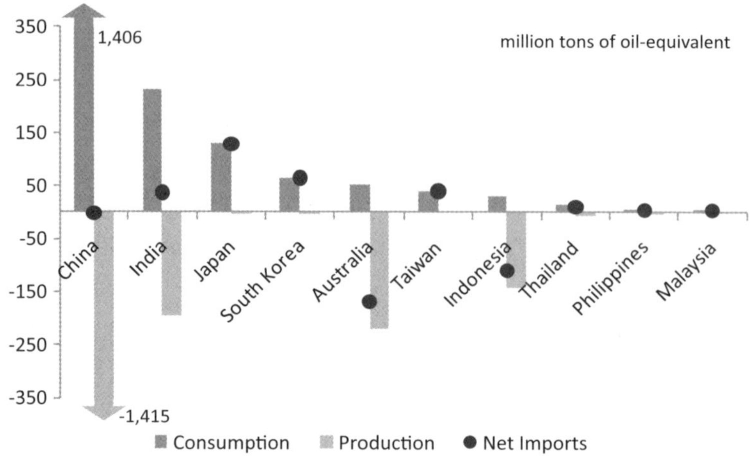

Figure 7 Asia Coal Consumption, Production and Net Imports 2008
Source: *BP Statistical Review of World Energy*, June 2009.

most abundant domestic resource.[22] Among Southeast Asian countries, Thailand, Malaysia and Singapore have a well-developed domestic gas infrastructure. However, Thailand's gas reserves are depleting, and just like Singapore, the country is planning for future LNG imports. The Philippines has found one major offshore gas field that now supplies three power plants. The biggest challenge the country faces is that most of its larger potential oil and gas deposits are located in deepwater areas, and thus costly and difficult to explore and develop. Therefore, Indonesia, Malaysia and in the future Myanmar and Vietnam will be the region's largest natural gas and LNG exporters, while Singapore, the Philippines and in the future Thailand will be importing the bulk of their gas from their neighbours.

China, India and Japan are the largest coal consumers in Asia (Figure 7). China is relatively self-sufficient in both thermal and coking coal, India imports higher-quality thermal and coking coal while Japan

[22] See National Energy Policy Council (NEPC), *Thailand's Energy Policy and Development Plan under the Administration of Prime Minister General Surayud Chulanont* (Bangkok: NEPC, 2006); and PTT Plc, *Annual Report 2008* (Bangkok: PTT Plc, 2009).

and South Korea are the region's biggest importers. The strong demand from Northeast Asia benefited Indonesia, the world's largest exporter of thermal coal, as well as Australia (the latter is the world's leading coking coal exporter).

Thailand, Malaysia and the Philippines are growing importers, while Vietnam may soon see its domestic coal production insufficient to cover demand from its growing electricity generation capacity.[23] This raises the question of long-term coal supply security. Despite the transport cost advantages, these countries are competing with Asia's largest coal importers, Japan, South Korea, India and Taiwan, for several international coal suppliers. There is only one regional supplier of coal, Indonesia. While China has sufficient coal deposits, the growing Chinese demand for lower-ash and lower-sulphur thermal and coking coal keeps producers focused on their own domestic markets, at least for the foreseeable future. Although Indonesia was the world's largest exporter of thermal coal for power generation in 2008, benefiting from the relatively low sulphur and ash content of its coal, Australia's production capacity is continuously increasing.[24] Indonesia's main challenge is the capacity and infrastructure constraints in coal mining, processing and transport infrastructure.

A summary of the oil, natural gas and coal sector potential in Southeast Asia is given in Table 1. In both natural gas and coal Southeast Asian exporters cannot afford to become complacent, since competition from Australia and the Middle East will intensify in future.

1.3.2 Demographics, Economic and Industrial Development, Geography and Population Distribution

Figure 8 shows that demand for primary energy is positively correlated with per capita income. This is consistent with studies conducted by various authors.[25] As an economy grows and develops its manufacturing and agricultural sectors, industrial fuel, electricity and feedstock consumption

[23] Barlow Jonker, *Coal Profile — Vietnam 2007* (Sydney: Barlow Jonker, 2007).

[24] EIA, *International Energy Outlook 2009*, p. 55.

[25] Smil, *Energy at the Crossroads: Global Perspectives and Uncertainties*, p. 66.

Table 1 Summary of Southeast Asian Oil, Gas and Coal Sector Potential

	Crude oil		Fuel products		Natural gas		Coal	
	Domestic	Trade	Domestic	Trade	Domestic	Trade	Domestic	Trade
Indonesia	Significant but gradually declining production	Net importer since 2008	Insufficient refinery capacity, growing demand	Importer of products, exporter of LPG and fuel oil	Growing demand from power sector	Global top 3 LNG exporters	Growing demand from power sector	#1 global exporter of thermal coal in 2008
Malaysia	Significant production	Net exporter	Insufficient refinery capacity	Net importer	Several gas-fired power plants	Global top 3 LNG exporters		Importer of thermal coal
Thailand	Small production	Import dependent	Established refinery sector	Net exporter of products	Currently self-sufficient, established pipeline grid	Some import from Myanmar, future LNG importer	Some local production	Net importer

(*Continued*)

Table 1 *(Continued)*

	Crude oil		Fuel products		Natural gas		Coal	
	Domestic	Trade	Domestic	Trade	Domestic	Trade	Domestic	Trade
Philippines		Import dependent	Growing domestic household demand, insufficient refinery capacity	Net importer	One major project	Future LNG imports?	Small production	Net importer
Singapore		Import dependent, large trading hub	Large-scale refining and trading hub	Net exporter of refined products, importer of fuel oil		Import from Indonesia, Malaysia, future LNG		
Brunei	Limited domestic demand	Net exporter		Import dependent	Limited domestic demand	Net exporter		

Source: Author's own analysis based on statistical data presented and discussed in previous sections.

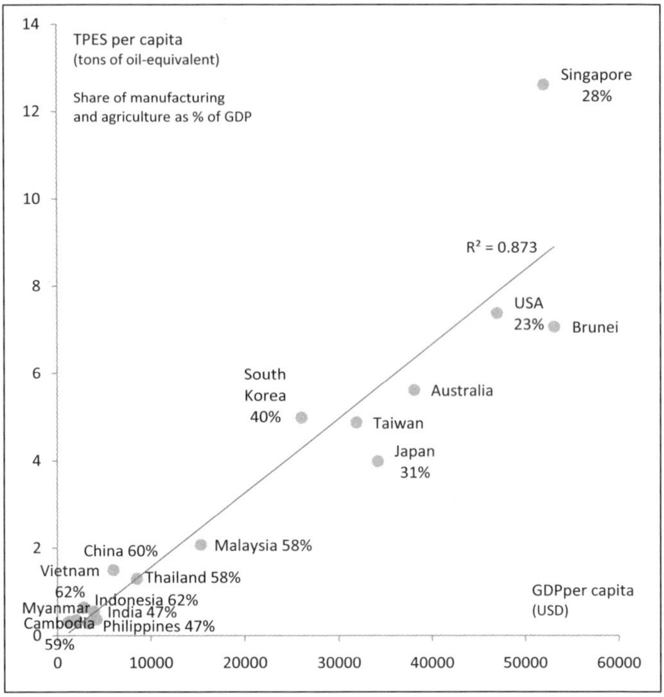

Figure 8 Total Primary Energy Supply (TPES) versus Per Capita Gross Domestic Product (GDP) 2008

Source: *BP Statistical Review of World Energy*, June 2009; EIA, *International Energy Outlook*, 2009; United Nations Development Programme (UNDP), *World Population by Country 1950–2050* (New York: UN Population Division, 2009); The World Bank, *World Development Report 2010: Development and Climate Change* (Washington, DC: The World Bank, 2010), pp. 384–385.

as well as increasing disposable income and household modernisation result in growing demand for primary energy.[26] In terms of energy intensity, i.e. the ratio between TPES and GDP per capita, Southeast Asian countries are in the middle of the range. This is compared to Japan, globally one of the most energy-efficient economies, and to the comparatively less efficient China. Figure 8 also shows that less-developed countries

---

[26] The agricultural sector would also require more energy inputs once modernisation takes place, i.e. through automation, high productivity seeds, large-scale irrigation and water pump technologies. For just one example, see Juan Pablo Rud, *Electricity Provision and Industrial Development: Evidence from India* (London: London School of Economics, 2009).

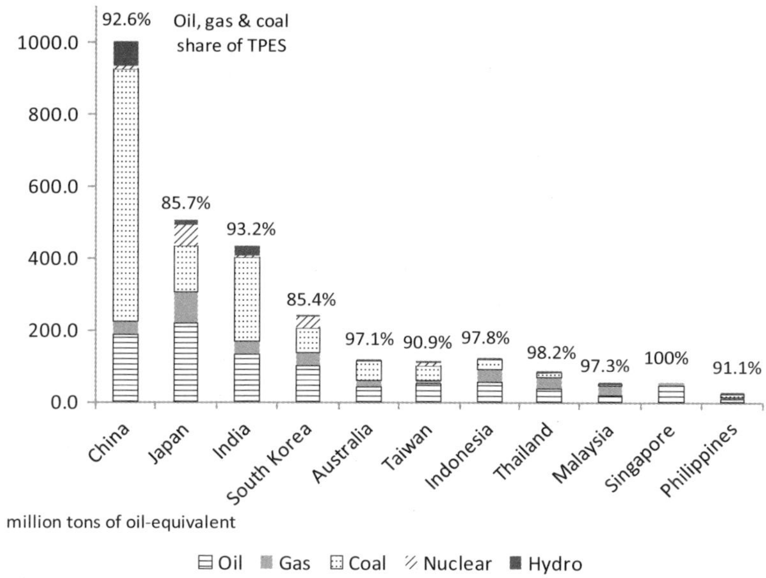

Figure 9 Primary Energy Consumption by Fuel Type

Source: *BP Statistical Review of World Energy*, June 2009; EIA, *International Energy Outlook*, 2009.

with high shares of manufacturing and agriculture as a percentage of GDP tend to consume higher levels of TPES and are more energy intensive. Singapore's and South Korea's high energy intensity are primarily driven by the significant fuel product re-exports from their refineries.

Figure 9 depicts the breakdown of primary energy consumption by fuel type and country. It can be seen that in Southeast Asia, Indonesia is the largest market for energy. The country consumes comparable amounts of oil, gas and coal, mainly for transportation fuel and electricity generation. The country's state-owned electricity company, PLN, is suffering from high fuel oil and diesel prices, and is aggressively diversifying away from liquid fuels to gas, coal and, in future, renewable energy. Thailand consumes its oil primarily for fuel and petrochemical feedstock. The Thai government has successfully implemented an industrialisation policy that optimises the use of the country's domestic gas potential.[27] Applications include the use

[27] NEPC, *Thailand's Energy Policy and Development Plan*.

of natural gas for light petrochemicals, i.e. ethylene, as well as liquid petroleum gas (LPG) and compressed natural gas for transport fuel.

The development of electricity generation capacity drives the demand for fuel oil, diesel, natural gas and coal. The low per capita GDP levels and geographic characteristics in Southeast Asia result in transport and electricity costs making up a relatively high percentage of household and industrial disposable income. The only exception is Singapore, a city state with one of the world's highest per capita GDP levels.

Furthermore, geography and population distribution significantly influence the need to transport feedstock and manufacturing goods, both to domestic markets and demand centres as well as to regional and international export markets. Indonesia and the Philippines, being two major archipelagos with hundreds of domestic demand centres, large and small, face more logistical constraints and challenges than the smaller, more localised countries like Malaysia, Thailand, Vietnam and of course Singapore. The transport of fuel products and energy feedstock to remote islands in Indonesia and the Philippines results in high costs of fuel and electricity generation, in turn affecting local competitiveness and development.

1.3.3 Refinery Sector and Fuel Distribution

Refinery facilities of an efficient scale require large capital investments. The sector's profitability margins are volatile and simple refining margins can at times be minimal.[28] Coupled with the importance of fuel prices for general transport, power generation and industrial consumption, as well as the sensitivity surrounding political decision making and regulatory capacity, the sector has historically seen a large role played by government-controlled state-owned enterprises (SOEs). While the role of SOEs in China, India and Southeast Asia has been important, each country in Asia has followed a distinct development path. Moreover, sector liberalisation

---

[28] "Simple" refineries typically comprise atmospheric, maybe vacuum distillation and produce lower-value-added fuel oil, heavier fractions and middle distillate products such as diesel and kerosene. In contrast, "complex" refineries (with a higher Nelson complexity index) include hydro and or catalytic crackers that reduce the proportion of heavier products and improve the yield of medium and light products such as gasoline, naphtha, diesel, kerosene and jet fuel.

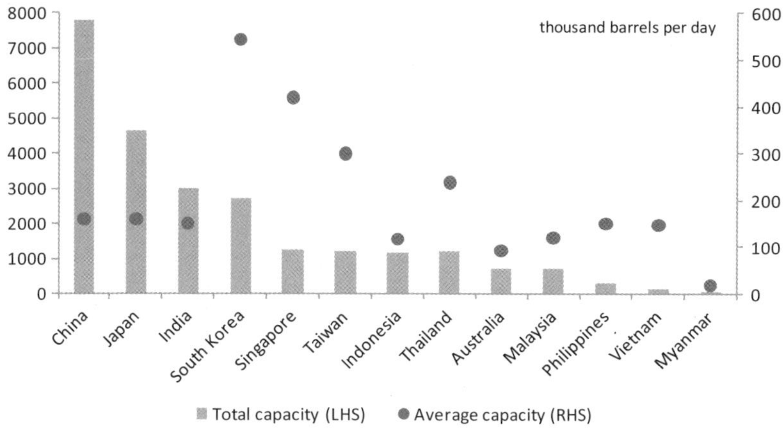

Figure 10 Total and Average Capacity of Asian Refineries

Source: *BP Statistical Review of World Energy*, June 2009; Worldwide Refinery Survey, *Oil & Gas Journal*, 2006.

during the last two decades has introduced private, domestic and/or foreign competition in many countries.

Figure 10 shows that, in terms of average capacities, the South Korean, Taiwanese, Singaporean and Thai refineries are the most efficient in the region. It should be noted that to achieve the minimum efficient scale, the consensus is that minimum crude processing capacities of about 150,000–200,000 barrels per day (bpd) are required. In comparison, in South Korea SK Corporation's Ulsan refinery has a capacity of 850,000 bpd, while Reliance Industries's two refineries will have a combined capacity of more than 1.2 million bpd. Thus it is apparent that the average refinery size in Southeast Asia, especially in Vietnam and Cambodia, is generally below the efficient scale.[29] In terms of the Nelson complexity index, which characterises the complexity, i.e. proportion of higher-value-added products such as lighter gasoline, diesel and kerosene, of a refinery's overall production output, only Singapore and Malaysia refineries achieve average or above average complexity in Southeast Asia (Figure 11). In terms of industry development, Indonesia's refinery and distribution sector has

[29] The exceptions include the larger-capacity Melaka II refinery in Malaysia, Petron refinery in the Philippines, and Cilacap and Balikpapan refineries in Indonesia.

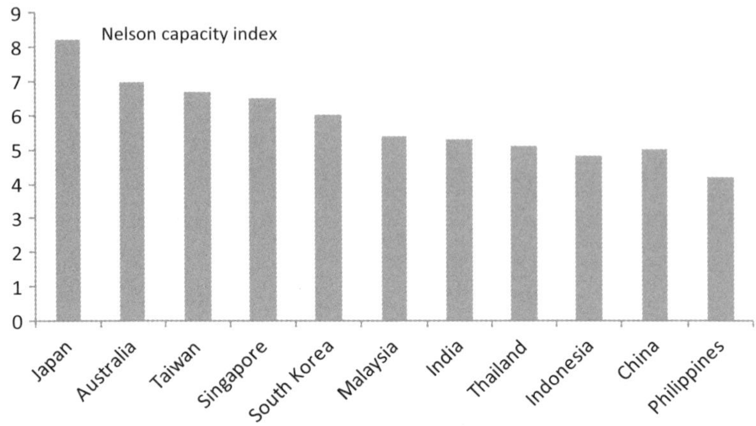

Figure 11 Refinery Nelson Complexity Index

Source: Worldwide Refinery Survey, *Oil & Gas Journal*, 2006.

been and still is dominated by its SOE, Pertamina. This is historical as well as a result of the low subsidised fuel prices, which renders it less profitable for private refiners. With the extreme increase in crude prices between 2005 and 2008, subsidies have been abolished for industrial customers and for high octane gasoline. SOE-multinational refinery joint ventures (JVs) are actively competing with refineries and retail stations owned by other multinationals in Malaysia, Thailand and the Philippines. In Vietnam and Myanmar the SOEs own one small refinery each.

In comparison, in China, most major refineries, following industry restructuring in the 1980s, are controlled and owned by state-owned Sinopec and CNPC-PetroChina.[30] China National Offshore Oil Corporation (CNOOC) is planning one major refinery and petrochemical complex, while the fourth government-controlled SOE, Sinochem, partially owns one Chinese-foreign refinery JV in Northern China. While historically Chinese refineries are small and less complex, the size and complexity of newly built and planned refineries, some of which are JVs between

[30] Although still government-controlled, all three major SOEs, Sinopec, PetroChina and CNOOC (just like ONGC in India, PTT Plc and several of its upstream and downstream subsidiaries and affiliates in Thailand), have been partially privatised and are now publicly listed on international stock exchanges.

Chinese SOEs and multinational oil majors, are increasing. This is driven by the need to refine heavier Middle Eastern crudes and to achieve better efficiencies of scale and scope.[31] In India, government-controlled Indian Oil Corporation, Bharat Petroleum and Hindustan Petroleum are facing strong competition from Reliance Industries, which has started operation of its first large-scale refinery and is about to double its capacity in a few years.[32] In South Korea and Japan, on the other hand, the refinery sector was developed with the help of large domestic business groups, some of which decided to form JVs with foreign strategic and financial partners.

It is noteworthy that there seems to be a rough correlation between the size and/or complexity of refinery sectors and the domestic crude availability across the countries. As Figure 12 indicates, countries that are heavily dependent on crude imports such as Japan, South Korea, Taiwan and

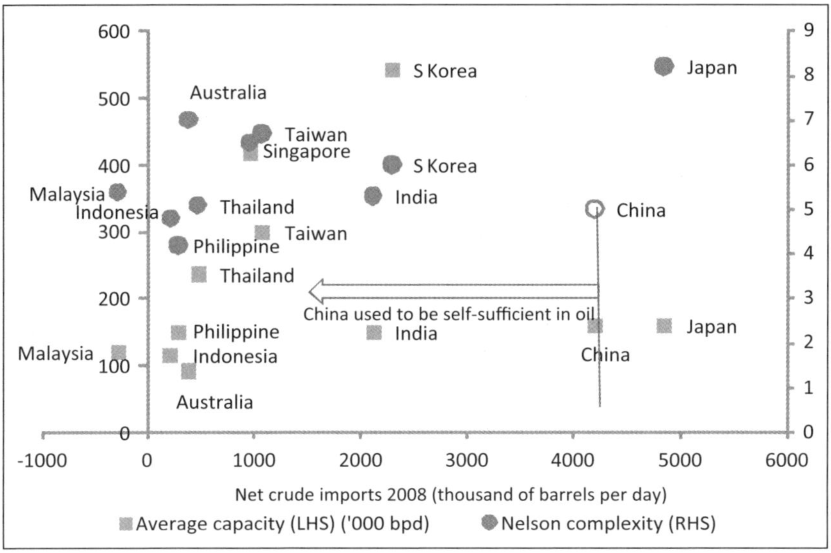

Figure 12 Refinery Size and Complexity versus Oil Import Dependency

Source: *BP Statistical Review of World Energy*, June 2009; EIA, *International Energy Outlook*, 2009; Worldwide Refinery Survey, *Oil & Gas Journal* 2006.

[31] Historically Chinese refineries have refined domestic crudes, e.g. from PetroChina's Daqing and Sinopec's Shenli fields, which are of lighter grades.

[32] Bharat and Hindustan are also publicly listed and thus partially privatised.

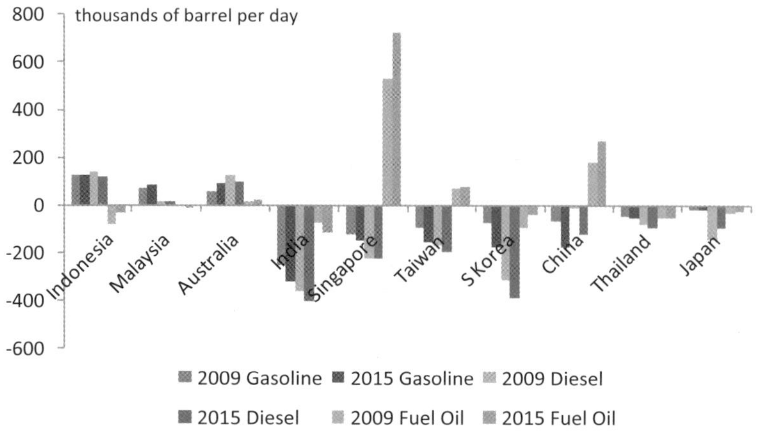

Figure 13 Net Imports of Diesel, Gasoline and Fuel Oil, 2009 and 2015
Source: Yamakawa, "Outlook of Products Supply & Demand in Asia by Country and by Product".

Thailand have generally built more complex and larger refineries, whereas countries that had (at least historically) a surplus of domestically produced crude oil, including China, Indonesia and Malaysia, have been "less forced" to do so. While it is conceivable that countries that import crude oil may have to focus more on achieving feedstock flexibility, margin protection and higher economies of scale and scope than countries with secure domestic crude availability, this apparent correlation may need to be examined in future studies.

The regional forecasted 2009 and 2015 net import figures for gasoline, diesel and fuel oil are depicted in Figure 13. In terms of domestic supply demand balance, only Indonesia and Malaysia need to import transport fuels. Given the lower average complexity of Indonesian refineries, however, the country is and will remain a net exporter of fuel oil for power generation and cracking. Thus, in principle the country could build additional cracking capacity to produce higher-value-added diesel, gasoline as well as kerosene and jet fuel. Singapore imports large volumes of fuel oil which its refineries crack into higher-value-added products, which it then exports Asia-wide.

After the closure of one refinery in the Philippines, the country is now relatively self-sufficient, while Vietnam and Myanmar are also forecasted to import lighter fuels until further refinery capacity comes on stream.

1.3.4 Energy and Fuel Subsidies

Governments in low-income countries frequently subsidise fuel and electricity prices. This is done to assist industrial consumers to become more competitive or to help domestic households improve their purchasing power. Figure 14 depicts the relationship between Asian countries' per capita GDP and domestic retail diesel and gasoline prices. While the logic of favouring low-income consumers renders subsidies in lower-income countries socially desirable, the correlation is far from perfect. The reason is that subsidy decisions are costly, especially in developing countries with constrained public budgets, and therefore politically controversial.[33]

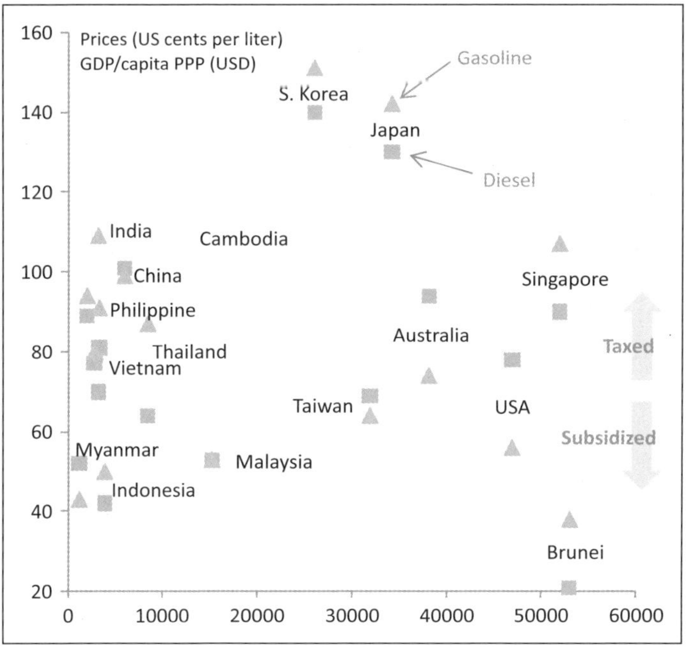

Figure 14 Per Capita GDP versus Diesel and Gasoline Retail Prices

Source: CIA, *The World Factbook* (2008 estimate); GTZ, *International Fuel Prices* 2009.

[33] The government soft budget constraint issue is a subject of extensive research. See for example János Kornai, "Ten Years After the Road to a Free Economy: The Author's Self Evaluation", in *Annual Bank Conference on Development Economics 2000*, edited by Boris Pleskovic and Nicolas Stern (Washington, DC: The World Bank, 2000), pp. 49–66.

Furthermore, retail price subsidies are often criticised for benefiting middle- and higher-income households as well, which runs counter to the original concept of redistribution.[34] It should be noted that retail prices in the United States are considered thresholds that allow average refiners to recover their crude price input and production cost, and still enjoy normal returns, while a 10% fuel tax is charged to help fund road and highway development and operational costs.[35] Thus, to make simplifying assumptions about general refining and distribution costs and margins, countries with retail prices above US levels generally extract taxes from consumers, while those below offer retail price subsidies. Figure 14 indicates that Indonesia, Malaysia and Brunei still offer fuel subsidies, while retail fuel prices in Thailand and the Philippines are generally in line with regional levels.[36]

Figure 15 depicts the development of diesel prices since 2000 across Asia. Interestingly, China, a country that has been criticised for maintaining below-market retail prices in the past, together with Vietnam and the Philippines, has allowed its fuel prices to gradually increase, somewhat in tandem with the oil price increase between 2005 and 2008. Chinese retail fuel prices are now higher than levels in the US. While Thailand and India also allowed their domestic retail prices to increase until late 2007, during the oil price peak in the first half of 2008, these countries "encouraged" their refineries to share at least part of the cost of subsidies and keep prices below market levels. Indonesia, on the other hand, has gradually phased out its fuel subsidies, with the only subsidies remaining for standard diesel, gasoline and kerosene for retail and households. As can be seen, consumers in the more advanced countries of Singapore, Australia, China, Japan and South Korea pay higher retail price levels as a result of higher levels of fuel taxes.

---

[34] See for example IEA, *Energy Policy Review of Indonesia* (Paris Cedex: IEA, 2008); from a theoretical point of view, optimum subsidies are most effective if directly targeted at the main beneficiaries, e.g. specific industries and sectors, low-income households. See Dani Rodrik, "Policy Targeting with Endogenous Distortions: Theory of Optimum Subsidy Revisited", *The Quarterly Journal of Economics*, 102 (1987), pp. 903–911.

[35] See GTZ, *International Fuel Prices 2009* (Eschborn: GTZ, 2008), pp. 3 and 64.

[36] See for example Neil Atkinson, *Subsidised Oil Prices: Are They Sustainable?* (Manila: Department of Energy, 2008).

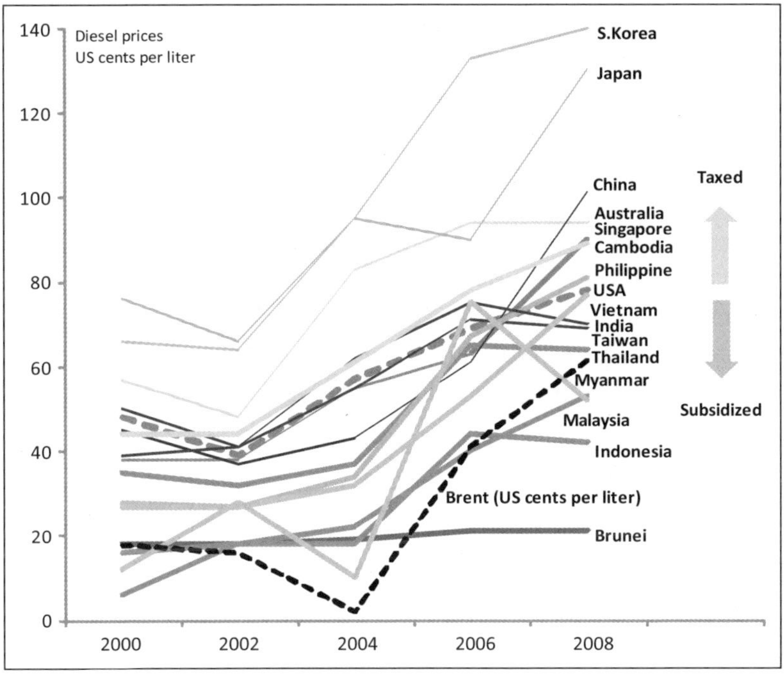

Figure 15 Diesel versus Crude Oil Prices 2000–2008

Source: GTZ, *International Fuel Prices 2009.*

1.3.5 Institutional History and Foundation

As many researchers have demonstrated, economic policies cannot be effective unless a country's socio-political, institutional history and foundation are all aligned. Taking the case of industrial policy, for example, the experience in East Asia has shown that industrial policy has been successful when those in power who have implemented the policy have either themselves wished for industrialisation to succeed, or have been forced to act in this way by the incentives generated by political (and socio-historical) institutions.[37]

[37] James A. Robinson, "Industrial Policy and Development: A Political Economy Perspective", *2009 World Bank ABCDE Conference* (Seoul: The World Bank, 2009).

The policies of Southeast Asian governments for their oil, gas and electricity industries have been shaped not only by each country's resources, i.e. energy feedstock endowments, geography and demographics, but by their institutional history and socio-political dynamics as well. Only the three examples of Indonesia, Thailand and Malaysia will be elaborated. Since the formulation and implementation of energy policy is a gradual, evolutionary process, the institutional backdrop and historical energy sector development significantly influence, and often limit, the flexibilities and policy options that each successive government in these countries are able to formulate and, in particular, implement. Here we mainly look at the institutional background in Indonesia, Thailand and Malaysia.

1.3.5.1 *Indonesia*

In Indonesia, Southeast Asia's oldest and largest oil- and gas-producing country, industry and regulation have evolved in several stages.[38] Starting in the early days of production sharing contracts (PSCs) about four decades ago, Indonesia became one of the developing world's most actively explored and developed oil and gas markets. In 1970 the government created a unified national oil and gas company (NOC), Pertamina, which was to be controlled by and managed under the Ministry of Mines and Energy. While the government, through Pertamina, continued to engage in various cooperation contracts (including PSCs) with domestic and foreign oil operators, Pertamina was not given sufficient freedom to develop independent and commercial operations.

This was only addressed following the 1997 Asian crisis, the 1998 fall of the Suharto government and the 1999 decentralisation laws. In 2001 the Indonesian government introduced a new oil and gas law that provided the basis for a more commercial and transparent industry and regulatory structure, independent upstream and downstream regulatory bodies and for a more commercially nimble, corporatised Pertamina.[39] In

[38] For a review of Indonesian energy policy, its oil, gas and coal industry and the country's need to prioritise institutional improvements such as policy coordination and decision making, regulatory independence and sustainable development, see IEA, *Energy Policy Review of Indonesia*.

[39] See IEA, *Energy Policy Review of Indonesia*, pp. 105–120.

the country's more fragmented mining sector, on the other hand, a new mining law was introduced in 2009, which introduced a more decentralised industry structure and licensing regulations.[40]

Among the problems facing the Indonesian oil, gas and mining sectors are the comparatively less attractive oil and gas fiscal regimes, a complex sector bureaucracy and the prevalence of corruption and special interest groups. Coupled with institutional weaknesses at the regional government level following the country's extensive 1999 decentralisation laws, this has resulted in a less conducive investment environment and declining oil and gas production in recent years.

1.3.5.2 *Thailand*

The structure, regulation and SOE privatisation policies in Thailand's gas and electricity sectors have been heavily debated along two dimensions. First, the proponents of market competition and unbundling (e.g. of the downstream gas and electricity sectors) have been pitted against the supporters of large, efficient and integrated "national champion" SOEs. Second, the proponents of (partial) SOE privatisation have been battling consumer groups and non-governmental organisations who see more disadvantages of privatisation.[41] Nevertheless, most parties seem to agree on the need to maximise the country's utilisation of its domestic natural resources for the production and distribution of transport fuel and electricity, with the ultimate goal of driving the country's industrialisation and development.

One successful case study is the privatisation of the Thai NOC, PTT Plc (formerly Petroleum Authority of Thailand). While the government of the

---

[40] For an overview of Indonesia's coal sector, see IEA, Energy Policy Review of Indonesia, pp. 153–169; for a discussion on the new Mining Law, also see Balbir Bhasin and Sivakumar Venkataramany, "Mining Law and Policy: Replacing the Contract of Work System in Indonesia", CEPMLP Research (2008), pp. 1–16, available from http://www.dundee.ac.uk/cepmlp/.

[41] On the debate about industry structure and privatisation, see Puree Sirasoontorn and John Quiggin, "The Political Economy of Privatization in the Thai Electricity Industry", *Journal of the Asia Pacific Economy*, 12, no. 3 (2007), pp. 403–419; for an overview of the Thai oil and gas sector policy and history, see NEPC, *Thailand's Energy Policy and Development Plan under the Administration*; and PTT Plc, *Annual Report 2008*.

late 1990s, working with the National Energy Policy Office (NEPO), preferred a more competitive, unbundled and privatised industry, following the change in government in 2001 and the PTT Plc initial public offering (IPO) the company maintained its quasi-monopoly of the gas transmission grid. As part of its restructuring and corporatisation in preparation for the privatisation, it was stipulated that PTT subsequently carve out and unbundle its gas transmission and distribution (T&D) business.

The country's downstream oil refinery and petrochemical industry was developed via the formation of JV companies between PTT's subsidiaries and domestic or multinational partners. Some of these JV affiliates have subsequently been consolidated and/or partially privatised via stock market listing. This strategy has helped PTT join Singapore and Malaysia as one of Southeast Asia's most efficient refinery and petrochemical locations.

Thailand's and PTT's main challenge is the potentially depleting domestic gas resources, and PTT's aggressive acquisition of domestic gas and oil reserves and engagement in regional and international oil and gas projects is part of their strategy to mitigate future dependence on imported gas.[42] PTT has also been mandated to plan a future LNG receiving terminal. On the renewable energy side, the government is pursuing fuel diversification strategies, particularly into biofuels and biomass, as well as hydro, wind and solar power.[43]

1.3.5.3 *Malaysia*

Malaysia, on the other hand, pursued an oil and gas strategy centred on Petroliam Nasional Berhad (PETRONAS). PETRONAS has evolved to

[42] For analysis of Thailand's future gas import dependence, energy security and fuel diversification plan, see Thanawat Nakawiro and Subhes C. Bhattacharyya, "High Gas Dependence for Power Generation in Thailand: The Vulnerability Analysis", *Energy Policy*, 35 (2007), pp. 3335–3346; and Thanawat Nakawiro, Subhes C. Bhattacharyya and Bundit Limmeechokchai, "Electricity Capacity Expansion in Thailand: An Analysis of Gas Dependence and Fuel Import Reliance", *Energy*, 33 (2008), pp. 712–723; for a summary of the Thai Ministry of Energy's alternative energy policy, see Areerat Yoohoon, "Low Carbon Economy", presented at the Asia Pacific Forum on Low Carbon Economy (Beijing: Department of Alternative Energy Development and Efficiency, Ministry of Energy of Kingdom of Thailand, 2009).

[43] Yoohoon, "Low Carbon Economy".

become one of Southeast Asia's most successful and internationally diverse oil and gas SOEs. It started when then Prime Minister Mahathir invited a group of professionals under the leadership of Hassan Marian to build a world-class oil and gas SOE.[44] PETRONAS top management was given full discretion to recruit and train the best oil and gas professionals and executives. More importantly, the government gave PETRONAS the authority and independence to conduct its business in the most technically and commercially optimal manner, with minimum government intervention. The company thus managed to become an international LNG player (the only one among Asian SOEs), an efficient and dominant domestic upstream and downstream operator, and an active explorer across Africa, Southeast and Central Asia, with growth strategies in the European and North American gas markets. Domestically, the 1974 Petroleum Development Act gave PETRONAS the exclusive rights to explore and develop prospective oil and gas fields, either by themselves or in cooperation with appropriate multinational or domestic industrial partners.

1.3.6 Upstream Fiscal Policy

Analysing the oil and gas sector competitiveness and investment attractiveness across countries requires a comparison of not only aggregate fiscal policy but also the relative reservoir, topographic, climate and market conditions.[45] Thus, comparing average, aggregate fiscal regimes only is a gross simplification. Moreover, there are a wide variety of different contract types, many of which combine elements of revenue royalty (e.g. first tranche petroleum and domestic market oil and gas) with profit and corporate taxations. These contractual elements interact with the aforementioned geographic, reservoir and market variables, resulting in a complex, country-specific risk and return profile which changes over time with the actual portfolio of projects in each country.

[44] See Fred R. von der Mehden and Al Troner, *PETRONAS: A National Oil Company with an International Vision* (Houston: The James A. Baker III Institute for Public Policy and Japan Petroleum Energy Center, Rice University, 2007).

[45] See Cambridge Energy Research Associates (CERA), *A Comparison of Fiscal Regimes* (Cambridge: CERA, 2007).

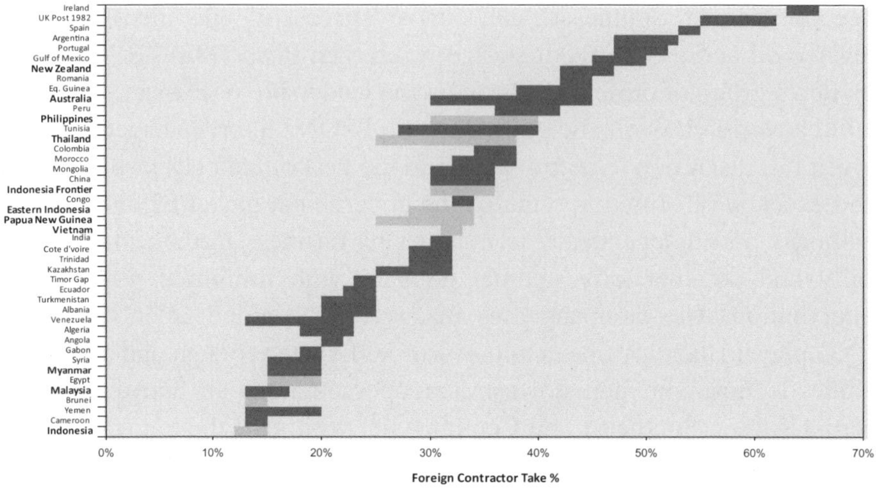

Figure 16 World Oil and Gas Fiscal Comparison

Source: IHS Energy Group.

Nevertheless a simple comparison of aggregate contractor versus government take as in Figure 16 could provide a rough idea of where Southeast Asian countries stand vis-à-vis the region's global competitors. The latter nowadays primarily comprise offshore West and North Africa, Central Asia and Brazil, as well as Australia.

Figure 16 shows that frontier (generally including gas) and eastern Indonesia, Vietnam, Thailand and the Philippines are on average in the middle of the worldwide fiscal regime scale. However, judging from recent large-scale discoveries in West and North Africa, and Brazil, as well as the investment activity involving large fields in Central Asia, these countries' lower or similar foreign contractor takes (e.g. about 20% in Angola and Gabon, 30% in Kazakhstan) seem to be over-compensated by these regions' higher reserves and resource potential. It is well known that the fiscal regime in Australia is comparatively more contractor-friendly, which coupled with the very large potential for gas and LNG and low political risk, makes it highly competitive.[46] Add to this the fact that future exploration activity in

[46] Upside potential exists for future LNG expansion in Australia, despite possible delays due to the financial crisis, forecasted medium-term over-supply and environmental regulation. See EIA, *International Energy Outlook 2009*, p. 43.

Southeast Asia, driven by both geological and local-political factors,[47] will focus increasingly on higher cost and risk offshore projects, it could be inferred that improving the contractor share may provide additional incentives for both foreign and domestic exploration activities.

On the coal side, the key competitors in international and regional export markets are Indonesia and Australia. Australia's coal royalty regimes, while varying across states, range from 3.5% in Southern Australia to around 7% in Queensland and New South Wales.[48] This is lower than the 13.5% level of central government royalties charged to Coal Contracts of Work (CCOW) and the 7% historical range for local concessions in Indonesia.[49]

1.3.7 Regional Summary

Each country in Southeast Asia faces a distinct combination of geographic, demographic, economic and socio-political characteristics that help determine the potential future of the country's energy sector. These determinants of energy policy are summarised in Table 2.

This creates unique opportunities and challenges for each country. While almost all countries in Southeast Asia are already or will become dependent on imported oil in the future, Indonesia, Malaysia, Brunei, as well as possibly Vietnam and Myanmar, could become future suppliers of gas to both regional and international LNG and pipeline gas consumers.

---

[47] For example, in Indonesia, lower-cost and -risk onshore basins, especially in western Indonesia, have been extensively explored, while large potential is still anticipated for offshore discoveries in eastern Indonesia. In terms of project implementation the current decentralised political environment has resulted in a multitude of local policy problems, e.g. bottlenecks in land clearing, the proliferation of local regulations, taxes and duties as well as social issues, which could be partially addressed in the case of offshore projects.

[48] Craig Bowie, *A Review of Mining Royalties in Australia* (Brisbane: Minter Ellison Lawyers, 2009). In Western Australia, the royalty for domestically supplied coal is AS$1/tonne, whereas in Queensland an additional 3% royalty is paid on the value of the coal price exceeding a price of AS$100/tonne, while in New South Wales royalties vary between 6.2% and 8.2% depending on project characteristics.

[49] Ministry of Energy and Mineral Resources (MEMR), Republic of Indonesia. Aggregate fiscal terms in Indonesia's coal and mining sector may also change with the introduction of the Mining Law and the increased devolution of power to local district and municipal governments.

Table 2 Summary of Determinants of Energy Policy

	Geography and resource endowment	Demographics	Economy and industry	Institutions
Indonesia: oil and fuel importer, gas and coal exporter, weaker institutions and investment climate	Large archipelago with more than 17,000 islands, rich agricultural soil Exporter of gas and coal, import dependent in crude and fuel products. High exploration potential in gas and coal, medium in oil, high potential for alternative energy including biofuels,[50] hydropower, geothermal, solar.	Lower per capita GDP (US$3,900). Population of 227 million.	Large agriculture sector, primarily labour-intensive industries and selected manufacturing (including basic chemicals, tires, plastics textiles), growing power and fuel demand, insufficient funds for pollution control and new energy technologies,[51] insufficient refinery capacity, subsidised retail gasoline and diesel. TPES including oil, gas, coal and hydro; geothermal and biofuels potential.	Established legal and regulatory base but weak enforcement, gradually stronger SOE, conflicts and special interests, difficult investment environment, less favorable fiscal regime.[52]

(*Continued*)

[50] Djatnika S. Puradinata, "The Biofuels Industry in Indonesia: Opportunities and Challenges", in Lugg and Hong, *Energy Essays on the Asia-Pacific Region*, pp. 42–55.

[51] See Hermawan and Sudarto P. Hadi, "Existing Sustainable (Renewable) Energy System in Indonesia", presented at The 2nd Joint International Conference on Sustainable Energy and Environment (SEE 2006) (Bangkok, 2006).

[52] IEA, *Energy Policy Review of Indonesia*, pp. 26–27 (Executive Summary), pp. 29–52 (Government Policy, Structure and Process).

Table 2 (*Continued*)

	Geography and resource endowment	Demographics	Economy and industry	Institutions
Malaysia: fuel importer, oil and gas exporter, strong SOE and established institutions	Peninsula, rich agricultural soil. Exporter of LNGs and oil, importer of fuel products and coal, top 2 producers of palm oil, high potential for alternative energy including biofuels, solar, hydropower.	Medium per capita GDP (US$15,300). Population of 27 million	Established agricultural and manufacturing sectors (latter including higher-value-added plastics, microelectronics), insufficient refinery capacity, subsidised fuel and gas. TPES including oil, gas, coal and hydro, biofuels.[53]	Established legal and regulatory base, strong NOC with international presence, incumbent political party.

(*Continued*)

[53] Also see Abdul Hamid Jafar, Abul Quasem Al-Amin and Chamburi Siwar, "Environmental Impact of Alternative Fuel Mix in Electricity Generation in Malaysia", *Renewable Energy*, 33 (2008), pp. 2229–2235.

Table 2 (*Continued*)

	Geography and resource endowment	Demographics	Economy and industry	Institutions
Thailand: oil importer, future gas imports, stronger SOE	Mid-size, Asian mainland, rich agricultural soil and offshore resources. Currently self-sufficient in gas, import dependent in oil, net importer of coal, future need for gas imports including LNG, potential for alternative energy including biofuels, wind, solar	Medium per capita GDP (US$8,500). Population of 67 million	Established mid-stage manufacturing sector, large agricultural sector (established exporter of agricultural products). TPES including gas, oil, coal and hydro.	Strong publicly-listed NOC, established legal and regulatory base, privatisation polemic.

(*Continued*)

Table 2 (*Continued*)

	Geography and resource endowment	Demographics	Economy and industry	Institutions
Philippines: oil, gas and coal importer, weaker institutions and investment climate	Mid-size archipelago with more than 7,000 islands, rich in agricultural soil. One major gas production site, import dependent in oil and coal, region's key geothermal power producer, offshore exploration potential, high potential for alternative energy including biofuels, solar, hydropower.	Lower per capita GDP (US$3,300). Population of 90 million	Established agricultural sector, lower manufacturing share (primarily labour-intensive industries), insufficient funds for pollution control. TPES including oil, coal, gas, hydro and geothermal.[54]	Established legal and regulatory base but weak enforcement, diversified multi-subsidiary NOC, unbundled electricity sector, conflicts and special interests, difficult investment environment.

(*Continued*)

[54] See Department of Energy (DOE), *2007 Philippine Energy Plan* (Manila: DOE, Republic of Philippines, 2007).

Table 2 (*Continued*)

	Geography and resource endowment	Demographics	Economy and industry	Institutions
Singapore: oil and gas importer, strong refining sector, strong institutions	Small island city-state Import dependent on crude oil and gas, large trader of crude and worldwide exporter of fuel products, lack of land area	High per capita income (US$52,000, 2nd after Brunei). Population of 4.6 million	Most advanced economy in Southeast Asia, high share of service sector, higher-tech industries (e.g. microelectronics, pharmaceuticals, diversified chemicals), among top 3 refining centres worldwide, ability to finance alternative energy projects and pollution control technologies. TPES including gas and oil.	Strong legal and regulatory base, unbundling and competition focus, privatised sector partially owned by government and multinationals, conducive investment environment.

(*Continued*)

Table 2 (*Continued*)

	Geography and resource endowment	Demographics	Economy and industry	Institutions
Vietnam: growth potential in oil and gas, future importer of coal, developing institutions	Mid-size, Asian mainland, rich agricultural soil and offshore resources Small historical production but significant growth potential and resource base in oil and gas, soon to become net importer of thermal coal potential for alternative energy including biofuels.	Low per capita GDP (US$2,800). Population of 87 million.	Growing domestic demand for oil and coal, potential for gas power plants, large agriculture sector, growing labour-intensive manufacturing base, insufficient refinery capacity, insufficient funds for pollution control. TPES including oil, coal, gas and hydro.	Developing sector regulation and legal base, attractive investment environment, growing NOC.
Brunei: limited domestic market, oil and gas exporter	Small land area in northern part of Borneo. Exporter of gas and oil, importer of fuel products.	Highest per capita GDP in the region (US$53,100). Population of 392 thousand.	Primarily services, significant fuel subsidies, limited domestic energy demand, mainly fuel and electricity for households	One NOC.

Additionally, Indonesia has the potential to maintain its role as one of the world's main exporters of thermal coal for power generation. Besides oil import dependence, the challenges of Indonesia and the Philippines are institutional in terms of attracting private domestic and foreign investment in the sector. Thailand and Malaysia must be prepared to start importing gas and particularly oil in the near future, while Vietnam, Myanmar and Cambodia have to further develop their oil and gas sector regulatory framework and institutions to maximise the potential of their nascent upstream sectors. Singapore, on the other hand, needs to continue its fuel diversification programme, not only involving oil and gas but also to function as a catalyst to the regional development of alternative energy technologies.

1.4 POLICY INSTRUMENTS AND NORMATIVE ENERGY POLICY GRID

In an ideal setting, governments would formulate their energy sector strategies and policies by taking into account the outlined geographic, demographic, socio-political and institutional characteristics. While each country will tailor its energy policy to work within the boundaries and constraints of its natural and economic characteristics, the socio-political and institutional backdrop is an essential factor to be considered in devising the policy implementation programme. To be successful, a holistic approach needs to be pursued comprising industrial and energy sector strategy and regulation, trade policy as well as fiscal and investment policy.

1.4.1 Industrial, Trade, Investment and Fiscal Policy

Based on the country's geography, demographics and institutional history, a government would select certain target industries that it wishes to develop and promote.[55] In the context of, for example, energy sector

[55] For a brief discussion on industrial policy, see for example Ha-Joon Chang, "Industrial Policy in East Asia: Lessons for Europe", presented at the EIB Conference in Economics and Finance (Luxembourg: European Investment Bank, 2006); on the complementarities between industry targeting and cluster strategies, see Douglas Woodward, "Porter's

and fossil fuel–based chemicals and plastics industry policy, this involves identifying the country's comparative advantage in domestically available feedstock and related factors of production. Following this, an analysis of scale economies, technology and product linkages with existing production facilities and related industries, transportation and infrastructure requirements, trade opportunities and thus production capacity objectives for each target industry can be undertaken.

A preliminary commercial feasibility study can thus be conducted, to assess the potential economic viability and attractiveness for these target industries. In addition to these industries' potential profitability, the country's overall investment climate and financing environment are very important. The aggregate industrial policy across target industries will form a basis for more accurate energy and fuel demand requirements, and the overall energy policy for the country. The government could then introduce appropriate fiscal policies, including taxations, royalties and/or subsidy instruments to support these target industries, and to assist in the provision of necessary energy infrastructure.[56] The latter would include upstream and downstream oil, gas and coal sector infrastructure and capacities.

1.4.2 Energy Sector Regulation and Policy

Once the country's short- and long-term energy, i.e. fuel and electricity, requirements have been established and the required infrastructure and capacities determined, the optimal energy sector regulation and policy grid must provide the most cost- and time-effective regulatory framework and institutional foundation to attract investment into the targeted energy sub-sectors. These include questions of industry structure and entry, i.e. competition versus (regulated) monopolies across sub-sectors and segments along the energy value chain, and fair yet sufficiently attractive and transparent

Cluster Strategy versus Industrial Targeting", presented at the ICIT Workshop (Orlando: ICIT, 2004).

[56] See for example Ridwan Rusli, "Time to Invest in Indonesian Oil and Gas Sector", *The Jakarta Post*, 30 June 2009.

tariff and pricing formulae.[57] To facilitate sector development and project implementation, competitive and transparent licensing and auction procedures as well as efficient cost monitoring and control mechanisms would be required. Many countries have also introduced independent regulatory agencies to oversee bidding, licensing and cost monitoring activities.

Furthermore, the question of investment incentives will depend on the inherent profitability of each sub-sector. While in general, notwithstanding exploration and construction risk, the upstream oil, gas and coal sectors are profitable, downstream refinery and distribution margins, especially in countries still maintaining residual fuel subsidies, can be quite volatile. Here the question of providing possible subsidies to either low-income households or to the refinery industry, as alternatives to historical retail price caps, must be weighed against the potential budget constraints of the government.[58]

Trade policy needs to be tailored to the specific conditions in the sector. While energy feedstock–importing countries may choose to pursue liberal trade policies and allow import competition, temporary tariff protection and export subsidies may help stimulate refinery sector investment.[59] However, in the latter case the problems of complacency, capture and possible retaliation have to be anticipated. In summary, each country's trade

[57] In the upstream oil and gas industry this often involves cost recovery, i.e. reimbursement schemes that would render strict cost monitoring and control mechanisms and independent, credible regulatory agencies necessary.

[58] For a discussion on the possible trade-off between subsidising and privatising (less-) profitable yet strategic industries, see Emmanuelle Auriol and Pierre M. Picard, "Infrastructure and Public Utilities Privatization in Developing Countries", *The World Bank Economic Review*, 23, no. 1 (2009), pp. 77–100. The model predicts that in developing countries with tight budget constraints, privatisation and price liberalisation may be optimal for low-profitability industries but suboptimal for more profitable industries. However, without a credible regulatory agency, regulation is better achieved by public ownership.

[59] Import protection is certainly a controversial policy, although there are examples where these could be successfully implemented if the disadvantages such as efficiency losses can be mitigated, for example through domestic competition and/or credible regulations. See for example Robinson, "Industrial Policy and Development: A Political Economy Perspective", pp. 3–4; and Paul R. Krugman, *Rethinking International Trade* (Cambridge: The MIT Press, 1990), pp. 113–114.

and investment policy, as well as financial sector regulation, must be sufficiently conducive to attract and finance investments in the sector.[60]

1.4.3 Role of SOEs and NOCs

Another question concerns the potential use of oil, gas and/or mining SOEs to promote or catalyse investment in these strategic sectors. Many studies have debated the potential trade-off between private sector governance and commercial efficiency versus the potentially less efficient, policy-tainted SOE performance.[61] One possible way to look at this question is to examine the trade-off between the government budget constraint and fiscal policy.[62] Simplistically, for example, one could distinguish between the upstream oil, gas and coal industry, which can be assumed to be quite a profitable business despite project-specific exploration risk, and the downstream processing and distribution infrastructure, which is mildly profitable, or at times unprofitable during industry down cycles.

For the profitable upstream sector, where royalty and taxation payments and the regulatory frameworks are relatively well established in mature oil and gas sectors such as in Indonesia, Malaysia and Thailand, it can be welfare enhancing to allow both SOE and private firms to manage and operate projects. This is despite the typical institutional challenges, information asymmetry and corruption problems.[63] On the other hand, in the

---

[60] See IEA, *Energy Policy Review of Indonesia*, pp. 53–62.

[61] For an extensive review see William Megginson and Jeffry Netter, "From State to Market: A Survey of Empirical Studies on Privatisation", *Journal of Economic Literature*, 39 (2001), pp. 321–389; Auriol and Picard in "Infrastructure and Public Utilities Privatization in Developing Countries" also predict that profitable industries in developing countries can either remain state-owned, or under well-functioning regulatory regimes, be privatised or outsourced to private operators who then pay appropriate taxes, royalty and or franchise fees to the government. This could potentially include upstream oil, gas and mining SOEs, which are generally quite profitable.

[62] See Auriol and Picard, "Infrastructure and Public Utilities Privatization in Developing Countries".

[63] For general discussions on the resource curse, institutional weakness and corruption, see Tim Hartford and Michael Klein, "Aid and the Resource Curse: How Can Aid Be Designed to Preserve Institutions?", *Public Policy for the Private Sector*, note no. 291 (Washington, DC: The World Bank, 2005); Thomas I. Palley, "Lifting the Natural Resource Curse",

mildly profitable downstream sector, where cost monitoring is equally difficult but regulatory and monitoring institutions are often less established than in the upstream sector, and considering the possibility of privatisations generating lower than expected proceeds,[64] SOEs could play a more major role in helping to facilitate investments.

This implies that in advanced economies with stronger and more transparent regulatory sectors and monitoring institutions, the question of SOE versus private sector investment is less relevant in the face of efficient taxation and more sophisticated regulatory regimes.[65] In the larger, more competitive and developed markets, the relative roles, strengths and capabilities of firms matter more than ownership. In addition, the potential profitability and attractiveness of the oil, gas and coal sectors for private sector investor entry and participation are determined more directly by the geological potential and the purchasing power of industrial and household consumers in these countries.

Foreign Service Journal, 80 (2004), pp. 54–61; Richard M. Auty, "The IMF Model and Resource-Abundant Transition Economies: Kazakhstan and Uzbekistan", UNU World Institute for Development Economics Research (UNU-WIDER), Working Papers 169 (UNU-WIDER, 1999); as well as IEA, *Energy Policy Review of Indonesia*, pp. 26–27 (Executive Summary), pp. 29–52 (Government Policy, Structure and Process). On vcorruption and institutional weakness in Southeast Asia, see Ari Kuncoro, "Corruption, Decentralisation and Democracy in Indonesia", *East Asian Economic Perspectives*, 17, no. 2 (2006), pp. 25–39; Iwan Azis, "Institutional Constraints and Multiple Equilibria in Decentralization", *RURDS*, 20, no. 1 (2008), pp. 26–27; Bhasin and Venkataramany, "Mining Law and Policy: Replacing the Contract of Work System in Indonesia"; Transparency International, *2008 Report on Revenue Transparency of Oil and Gas Companies* (Berlin: Transparency International, 2008).

[64] Compared to upstream oil and gas companies, refinery privatisations are often more difficult to execute and generate comparatively lower proceeds. For more general empirical evidence on under-valued privatisations see Nancy Birdsall and John R. Nellis, "Winners and Losers: Assessing the Distributional Impact of Privatization", Working Paper 6 (Centre for Global Development, 2002); John R. Nellis, "Privatisation in Developing Countries: A Summary Assessment", *SAIS Review*, 27, no. 2 (2007), pp. 3–29.

[65] Competition and credible regulation reduces information asymmetry between (public or private) company management on one hand and regulatory agency fiscal offices on the other. Auriol and Picard in "Infrastructure and Public Utilities Privatization in Developing Countries" indicate that even duopoly competition helps to reduce information asymmetry (at least through the "benchmark competition"-type effect).

1.4.4 Normative Energy Policy Grid

In principle, a normative energy policy makes use of a combination of industrial, energy sector, trade, investment and fiscal instruments to maximise investment incentives in the relevant energy sub-sectors. In addition, an appropriate regulatory framework must be in place to induce and regulate state-owned and private investors and energy operators to make their investments and operate in a socially and environmentally sustainable manner. Table 3 summarises recommended normative policy grids for countries with a distinct set of energy sector challenges. It should be noted that this framework can also be extended to address the objectives of environmental sustainability and political security.

The principles outlined are as follows:

- **Net exporters of oil, gas and coal:** Target industrial sectors that maximise the consumption of indigenous commodities, while still engaging in fuel and energy efficiency programmes to allow maximum export volumes and revenues. Promote competition between private domestic and foreign and SOE producers, as well as open trade policies, to ensure sector-wide efficiency. Offer competitive fiscal and investment regimes.
- **Net importers of crude and fuel products:** Focus on less energy- and fuel-intensive manufacturing and agricultural sectors. Diversify power generation to other available feedstock, i.e. gas and coal, light petrochemicals using gas and condensate, and renewables. Diversify crude import sources and contracts, including spot versus long-term, with the appropriate use of hedging arrangements. In countries with sufficiently high demand for additional refinery capacity, build complex and efficient-scale refineries, as well as flexible oil and fuel product storage and transport infrastructure, to allow distribution flexibility and mitigate margin volatilities. Consider offering fiscal incentives to build and/or expand refinery and distribution infrastructure, coupled if necessary with temporary import protection. On the other hand, for countries with insufficient incremental domestic demand, the decision whether to build new refinery capacity should depend on the state of refinery technology and infrastructure. For Thailand and Singapore, the experience base and export market presence could make it feasible to expand capacity further.

Table 3 Summary of Policy Instruments and Normative Policy Grids

	Industrial policy	Energy sector regulation and policy	Trade and investment policy	Fiscal policy	Relevant countries
Net exporter of oil, gas and or coal	Focus on sectors that add value to domestic resources (e.g. chemicals, plastics); fuel efficiency and conservation programmes to increase commodity export potential	Promote private and SOE competition, direct power generation sector towards long-term self-sufficiency and fuel diversification, sector-wide efficiency programmes	Promote competition through open trade, invite foreign investment	Attract domestic and multinational investment through competitive royalty and tax regime	Indonesia (gas, coal) Malaysia (oil, gas) Vietnam, Brunei (oil, gas)

(Continued)

Table 3 *(Continued)*

	Industrial policy	Energy sector regulation and policy	Trade and investment policy	Fiscal policy	Relevant countries
Net importer of crude	Focus on agriculture and less oil-intensive manufacturing industries Fuel efficiency and conservation programmes Diversify to available feedstock, e.g. gas and condensate for basic petrochemicals[66]	Maximise exploration and growth potential for domestic production (if any); balance spot and long-term import supply contracts Diversify power generation to available domestic feedstock (i.e. gas and/or coal) Build large and complex refineries, efficient crude storage and transport infrastructure	Ensure open and competitive crude import markets	A. If there is domestic resource and production growth potential: competitive royalty and tax regime, offer fiscal incentives for SOEs (and possibly multinational investor-SOE JVs) B. If no or insufficient domestic resources: support crude importers Encourage SOE or private investment in crude storage and transport infrastructure	A. Indonesia, Thailand, Singapore, Malaysia (future) B. Philippines

(Continued)

[66] For countries with depleting crude but large gas and coal reserves, diversifying to gas and condensate as feedstocks for basic petrochemical production are feasible alternatives. However, the use of coal for chemical and synthetic oil production is technologically possible but economically not yet feasible. A good example is Sasol of South Africa.

Table 3 (*Continued*)

	Industrial policy	Energy sector regulation and policy	Trade and investment policy	Fiscal policy	Relevant countries
Net importer of fuel products	Fuel efficiency and conservation programmes Expand or build refinery capacity with scale and complexity (provided majority of output can be consumed domestically)	Balanced spot and long-term fuel import supply contracts (incl. crude if to be imported for new refining capacity) Support SOE and private investment in refinery sector, incl. product export contracts (to gain scale economies through larger refinery capacity), and/or crude import or partnership contracts; build efficient fuel product storage, transport and distribution infrastructure	A. Temporary fuel import limitations to support new refinery B. Open trade if require continued product imports	Potential subsidy for refinery and infrastructure investments by SOE and/or private investors Support SOE or private investment in product storage and transport infrastructure	Indonesia, Malaysia, Vietnam

(*Continued*)

Table 3 *(Continued)*

	Industrial policy	Energy sector regulation and policy	Trade and investment policy	Fiscal policy	Relevant countries
Net importer of gas	Diversify fuel mix for industrial consumers and power generation, but keep share of gas given environmental advantages	Sign long-term supply contracts for LNG and pipeline gas (pipeline more vulnerable and less flexible than LNG but cheaper for limited volumes) Optimise network and regulation of pipeline grid to improve domestic transport efficiency	Mix of bilateral LNG and pipeline gas contracts	Support SOE or private investment in gas import pipelines and LNG receiving terminal, as well as domestic pipeline network	
Net importer of coal		Cater fuel mix for power generation depending on availability of oil and/or gas: if domestic or import gas feasible, reduce future dependence on coal for environmental reasons		Support SOE or private investment in coal handling, processing, storage and transport infrastructure	

Source: Author's own analysis

- **Net importer of gas:** In Southeast Asia this group overlaps with oil-importing countries, i.e. the Philippines, Singapore and potentially in the future Thailand. Diversify between LNG and pipeline imports. Negotiate long-term LNG supply contracts with at least two regional and/or international suppliers. Import pipeline gas from Malaysia and Indonesia. Build an efficient domestic pipeline network to reduce local transmission and distribution costs with appropriate sector regulation based on market size and the capabilities of incumbents and potential entrants.
- **Net importer of coal:** Depending on domestic availability, direct the fuel mix for power generation towards environmentally cleaner and regionally available gas and, possibly in a smaller proportion, oil.

Based on this analysis we anticipate that countries with more established regulatory and legal infrastructure, and more conducive financial and lending markets, will develop a more sophisticated private energy sector. In such an environment, oil-importing countries will build large and complex refineries. On the other hand, lower-income countries with relatively weaker and less independent regulatory institutions may opt to promote stronger state-owned oil, gas and mining companies. Having stronger SOEs would also allow governments in low-income countries, to the extent they have the financial resources or are net feedstock exporters, to potentially offer fiscal assistance in a targeted manner.

1.5 ANALYSIS OF ENERGY POLICY IN SOUTHEAST ASIA

Numerous studies have reported on energy policy across Asia. The instruments of energy policy examined in this chapter include industry regulation, trade and fiscal policy. In addition to strategy and policy, the observed institutional dynamics and implementation track record is analysed as well.

1.5.1 Observed Policy Grid

Table 4 depicts the summarised energy policy grid for the region. The main focus of analysis will be on the observed policy grids in Indonesia,

Table 4 Comparative Overview of Observed Energy Policy Grids

	Indonesia	Malaysia	Thailand	Philippines	Singapore	Vietnam
Industrial policy and energy strategy						
Target industries	Agriculture, labor-intensive industries, petrochemicals, tires	Agriculture, mid-level manufacturing, palm oil	Refinery and petrochemical, plastics, automobile	Agriculture, labour-intensive manufacturing	High value added fuel exports, technology-intensive manufacturing	Agriculture, labor-intensive industries
Fuel mix for power generation	Approx. 59% coal, 26% gas, rest hydro, geothermal and fuel oil (2010E)	40% gas, 30% hydro, 29% coal (government target 2020)	68% gas, 12% coal, 2% liquids, 15% hydro (2011)	32% gas, 26% coal, 18% geothermal, 16% hydro, 8% liquids (2008)	89% gas, 8% liquids, 3% waste incineration (2020E)	28% hydro, 27% gas, 30% coal, 8% imported electricity, 3% renewable, 3% nuclear (2020 plan)

(*Continued*)

Table 4 (*Continued*)

	Indonesia	Malaysia	Thailand	Philippines	Singapore	Vietnam
Stated alternative energy strategy	Reduce oil share of TPES from 40% down to 20%, replace with renewable (biofuels, hydro/solar/wind, geothermal) 5% of TPES from biofuels by 2025 (direct or max. 10% blended into gasoline and diesel), grow hydropower			Alternative Fuels Program incl. biodiesel, bio ethanol, gas-based vehicles and autogas	Waste-to-energy, energy technology industry incl. R&D (solar, biofuels, fuel cells), trading (LNG, biofuels, carbon credits) Comprehensive energy efficiency program	Growth in hydropower, renewables (wind, solar, geothermal, biomass, etc.) and higher share of nuclear

(*Continued*)

Table 4 (*Continued*)

	Indonesia	Malaysia	Thailand	Philippines	Singapore	Vietnam
Energy sector structure and regulation						
Upstream oil and gas	Competitive, NOC (Pertamina) and private domestic, foreign, JVs	NOC (Petronas) dominates, directly or via JVs with multinationals	NOC (PTT) dominates, directly or via JVs with multinationals	PNOC JV with Shell, Chevron	n/a	NOC (PetroVietnam) plus JVs
Coal	Competitive, private and SOE (Bukit Asam, Aneka Tambang)	Industrial imports	Banpu and industrial importers	Competitive, private and SOE	n/a	Vinacomin plus JVs
Refinery	Integrated SOE plus 1 affiliate	Integrated SOE plus SOE JVs and multinational competition	4 SOE JVs plus multinational competition	SOE JV plus 1 multinational	Government-multinational JVs	Integrated NOC (PetroVietnam)
Gas T&D	Duopoly SOE & independents, bilateral off-take contracts	Monopoly SOE, bilateral off-take	Third-party access, dominant SOE, bilateral off take	SOE-multinational JV, bilateral off take	Unbundled, competition	PetroVietnam.

(*Continued*)

Table 4 (*Continued*)

	Indonesia	Malaysia	Thailand	Philippines	Singapore	Vietnam
Electricity	Single buyer integrated SOE, private generation with long-term contracts	Single buyer integrated SOE, competitive generation with long-term contracts	Single buyer integrated SOE, competitive generation with long-term contracts	Unbundled, competitive generation sector	Unbundled, competition on all levels	Single buyer integrated SOE
Fuel distribution	Competitive for industrial and unsubsidised household gasoline, subsidised retail fuels	Competitive, subsidised prices	Competitive, market pricing (at times, implicit pricing guidance)	Competitive, market pricing (at times, implicit pricing guidance)	Competitive	Competitive, subsidised prices
Petrochemical	SOE & competition	SOE and JVs	SOE JVs & independent	Independent	Independent JVs	SOE

(*Continued*)

Table 4 (*Continued*)

	Indonesia	Malaysia	Thailand	Philippines	Singapore	Vietnam
Institutional framework						
Regulatory body	Independent upstream (BP Migas) and downstream (BPH) regulators	Ministry of Energy, Water and Communications as regulator of oil, gas, power sectors	Independent Energy Regulatory Commission (ERC) for electricity and gas	Ministries (oil & gas, mining) Independent Energy Regulatory Commission (ERC) for electricity	Independent Energy Market Authority (EMA)	Non-independent (ministries)
Policy	Ministry of Energy and Mineral Resources (MEMR)	Policies set by planning units reporting to Prime Minister	Policies by Energy Policy and Planning Office (EPPO) under Ministry of Energy	Department of Energy (DOE) for oil and gas Mines and Geosciences Bureau under Department of Environment and Natural Resources (DENR)	Ministry of Industry	Ministry of Industry

(*Continued*)

Table 4 (*Continued*)

	Indonesia	Malaysia	Thailand	Philippines	Singapore	Vietnam
State-owned enterprises (SOE)	NOC Pertamina, state-owned gas company PG N, state-owned mining Bukit Asam and Aneka Tambang, private competition in all sectors Electricity SOE (PLN), private competition in generation	Integrated NOC (Petronas) exclusivity for all upstream oil and gas projects (own and JV) Electricity SOE (Tenaga), private competition in generation	Listed NOC PTT Plc, with upstream and downstream affiliates and JVs (many publicly listed) Electricity SOE EGAT, private competition in generation	Oil, gas and coal holding SOE (PNOC), listed business-level JVs with private firms, private competition Electricity SOE Napocor, unbundled privatisation plus competition in all segments	Publicly listed electricity T&D, privatised gencos, unbundled privatisation plus competition in all segments	Integrated NOC (PetroVietnam) exclusivity for all upstream oil and gas, Vinacomin for mining projects (own and JV)

(*Continued*)

Table 4 (*Continued*)

	Indonesia	Malaysia	Thailand	Philippines	Singapore	Vietnam
Fiscal regime	PSC with revenue royalty, profit tax and cost recovery Aggregate roughly 30% post-tax contractor take Local government ownership, levies and duties	PSC with revenue royalty, profit tax and cost recovery 15–25% local ownership rights through SOE	8.8–12.5% revenue royalty, 50% profit tax, no cost recovery	PSC with revenue royalty, profit tax and cost recovery Aggregate roughly 40% post-tax contractor take	Only downstream sectors available: corporate taxes	PSC with revenue royalty, profit tax and cost recovery 20–30% SOE ownership (frontier), 40–50% SOE (standard)
Trade policy	Open trade, some Domestic Market Obligation (DMO)	Open trade, some DMO			Open markets	

(*Continued*)

Table 4 (*Continued*)

	Indonesia	Malaysia	Thailand	Philippines	Singapore	Vietnam
Implementation outcome						
Investment activity and industry develop-ment	Insufficient investment SOEs and private	Petronas plus multinationals	PTT dominant	Insufficient PNOC and multinationals	Multinationals and JVs of government-linked firms Active promoter role of government	Active explora-tion and development PetroVietnam, Vinacomin and multinationals
Institutional challenges	Regulatory structure and procedures, decentralization and local autonomy	Continued dominance of NOC	Dominance of NOC along the gas chain	Limited domestic economy of scale	Competition from regional NOCs and SOEs	Newly developing SOEs and sector infrastructures

Thailand and Malaysia. Other selected Southeast Asian countries are introduced for comparison purposes, given the somewhat comparable development and growth trajectory of their economies and energy sectors, and their unique positions as competing importers of oil, gas and coal in Asia.

1.5.2 Normative versus Observed Energy Policy Grids

1.5.2.1 *Industrial Policy and Energy Strategy*

As summarised in Table 4, the industrial mix and general energy strategy, i.e. target fuel mix for power generation and primary energy mix of Southeast Asian countries, are in line with the logic outlined before. All governments target fuel diversification programmes based on each country's energy feedstock availability, and encourage the development of renewable energy sources. All countries are reducing the share of liquid fuels, i.e. fuel oil and diesel feedstock, for power generation. Indonesia is aiming for a diversified mix of TPES between gas (30%), coal (33%), oil (20%) and renewable energy (17%) by 2025. Malaysia is focusing on gas, hydro and coal for power generation, Thailand on gas, coal and hydro. The Philippines on the other hand is successfully establishing its share of hydro and geothermal power, and is promoting biodiesel and bioethanol. Thailand and Singapore, being the region's main future importers of gas, are actively planning LNG import terminals and pursuing new contracts to diversify their gas base. Last but not least, Singapore is leveraging on its location and legal-commercial infrastructure and pursuing a technology-intensive strategy to become a major player in energy trading and in the research and development (R&D) of future energy technologies.

1.5.2.2 *Industry Structure and Sector Regulation*

The upstream oil and gas sectors generally have an established regulatory framework. The role of SOEs, i.e. the respective NOCs in Malaysia, Thailand and Vietnam, is by design still dominant, while the market is quite competitive for the NOCs in Indonesia and the Philippines. The downstream oil sector is generally competitive,

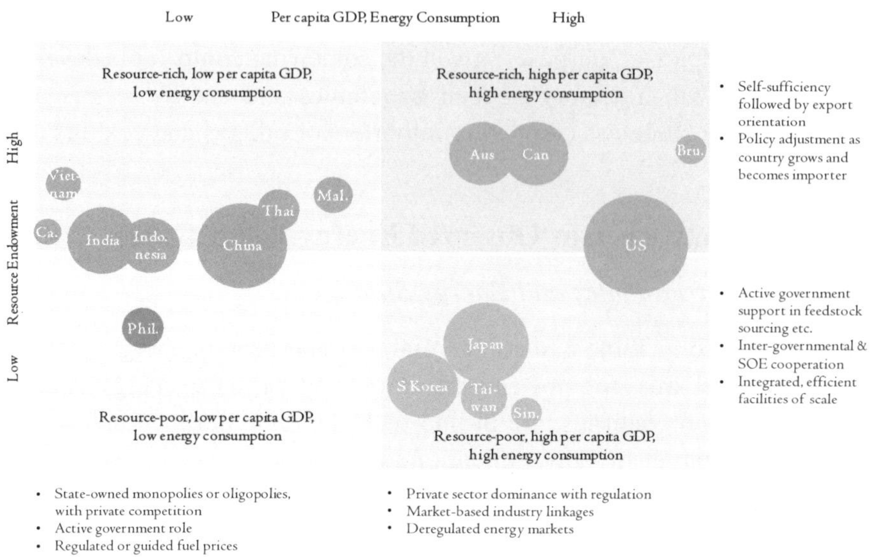

Figure 17 Country Comparison and Normative Energy Sector Policies

although the scale and ability to produce higher-value-added products is still limited in certain countries. The downstream gas sector, however, is still rather dominated by NOCs and SOEs in the three countries. Figure 17 summarises selected structural features of the oil and gas industry across Asia, while categorising each country within the appropriate resource endowment and per capita income quartile. Per capita income is assumed to serve as a very rough proxy for the government budget constraints, the degree of sophistication of domestic regulatory and monitoring institutions,[67] as well as the technology and funding capability of energy sector firms.

As anticipated, higher-income countries, such as Japan, South Korea and the US, with stronger regulatory and monitoring institutions see a

[67] This is a very crude way of characterising the region's regulatory institutions, given the institutional and historical differences. Nevertheless, despite these monitoring agencies being either independent (in Indonesia), under the NOC (Malaysia) or the Ministry (Thailand), the upstream oil and gas sectors are generally relatively well regulated.

larger participation of private sector companies as well as deregulated, competitive energy markets, while lower-income countries often require a comparatively larger role of sector SOEs, i.e. NOCs and the government. Additionally, countries constrained in natural resources, such as Japan, South Korea and Taiwan, which have the capability to build large-scale processing facilities, are faring much better than lower-income import-dependent economies, such as the Philippines, with limited ability to fund large-scale projects. Finally, low-income countries, especially those with strong public finances, such as China, or those that enjoy revenues from their commodity exports, such as Indonesia, Malaysia and Brunei, subsidise their fuel consumers. Thailand and the Philippines, on the other hand, allow market-based fuel pricing combined with "informal price management" during periods of very high oil prices.

1.5.2.3 *Trade, Investment and Fiscal Policy*

In general, policy is geared towards open trade and foreign competition in the downstream oil sector. The upstream oil, gas and coal sectors are open to foreign investors, although in Malaysia and Thailand the NOCs play a much more dominant role and participate in most projects in their country through the use of NOC-multinational JVs. In general this sector has been funded commercially, with minimal subsidies except for the aforementioned fuel distribution in selected countries. Investment laws are in place, dividend repatriation is generally free. The institutional challenges that somewhat mitigate foreign investor appetite in Indonesia and the Philippines are covered in the next section.

1.5.3 Implementation and Institutional Challenges

The main problem facing the region's governments is not so much the lack of coherent energy strategies and policies. The question seems to be whether or not the governments in a few countries could incentivise their current and future industry participants to take the risk and invest in frontier and new basin exploration and development projects, as well as in the required infrastructure and facilities.

In Indonesia this means that the oil and gas sector regulatory and monitoring procedures, and the tax reimbursements, must be processed more efficiently. Moreover, moves by special interest groups have delayed a number of upstream and downstream projects. Due to the lack of transparency and coherence in investment regimes, tendering and approval processes in the oil, gas, coal and general mining industries require better coordination between central and local governments and various parts of the government, in terms of cost recovery calculations, land clearing, local regulations, levies and duties.[68]

At the same time, better coordination amongst the region's NOCs and governments, beyond the usual rhetoric, could be advantageous. In Malaysia and Thailand, notwithstanding Petronas' and PTT's successes both domestically and internationally, their continued dominance in the domestic market make it more difficult for domestic players to develop their full technical and commercial potential. While in Malaysia this is observed both in the upstream and downstream oil and gas sectors, in Thailand this is more pronounced along the upstream and midstream gas chains. Singapore has build a highly efficient infrastructure that caters to its open market policy, but faces the growing threat of regional competition.

Moreover, in the resource industries in the Philippines, Vietnam and Cambodia the regulatory regimes, sector infrastructure and SOEs are still being developed. On the one hand, in the Philippines, this is driven by the country's limited resource prospectivity and import dependence. On the other hand, sector liberalization and exploration activity in Vietnam, and particularly in Cambodia, started accelerating more recently. For example, given that Malaysia imports increasing volumes of fuel products annually, PETRONAS and Pertamina could theoretically team up in building a new, more complex refinery capacity. Pertamina could enjoy enhanced credit by signing a product supply off-take agreement with

---

[68] See IEA, *Energy Policy Review of Indonesia*, pp. 26–27 (Executive Summary), pp. 29–52 (Government Policy, Structure and Process); as well as Ana Duek and Ridwan Rusli, "Democratised Indonesia and the 'Big Bang' Decentralization: The Impact on the Natural Resource Industry", unpublished research article, (Asia Competitiveness Institute, 2009).

PETRONAS for part of its output, while the partners could jointly negotiate long-term crude purchase agreements with appropriate suppliers. The same logic applies to the possible joint construction of an LNG terminal in Java, which could purchase domestic, Malaysian and/or international LNG to diversify Java's gas supply. This reduces its future dependence on the Sumatra-Java and possible future Kalimantan-Java pipelines.

1.5.4 More on Institutional Weaknesses

Several international surveys might help in better identifying the institutional factors that hamper the implementation of energy policies. First, a comparison of the quality of government institutions across countries is regularly undertaken by the World Bank through its Worldwide Governance Indicators (WGI) studies.[69] Figure 18 summarises the WGI for government effectiveness, regulatory quality, rule of law and corruption control. Although these are not sector-specific studies the results are indicative and consistent with the qualitative observations in the Southeast Asian oil, gas and mining sectors.

Figure 18 shows that within Southeast Asia, the legal and regulatory institutions are comparatively weak across the various criteria examined in Indonesia, Vietnam, Cambodia and particularly Myanmar. Rule of law is weak in the Philippines, and corruption control is still a problem in Thailand. As expected, Singapore scores very highly, in line with its worldwide reputation as an efficient and corruption-free government.

Second, in terms of ease of doing business, compared with over 180 countries worldwide Southeast Asian countries rank number 12 (Thailand), 23 (Malaysia), 93 (Vietnam), 122 (Indonesia), 144 (Philippines), 145 (Cambodia), while Singapore ranks number one.[70] This World Bank survey compares countries using multiple criteria such as ease of company

[69] Daniel Kaufmann, Aart Kraay and Massimo Mastruzzi, "Governance Matters VIII: Aggregate and Individual Governance Indicators 1996–2008", World Bank Policy Research Working Paper 4978 (The World Bank, 2009).

[70] See World Bank, *Doing Business 2010: Measuring Business Regulations* (Washington, DC: The World Bank Group, 2009).

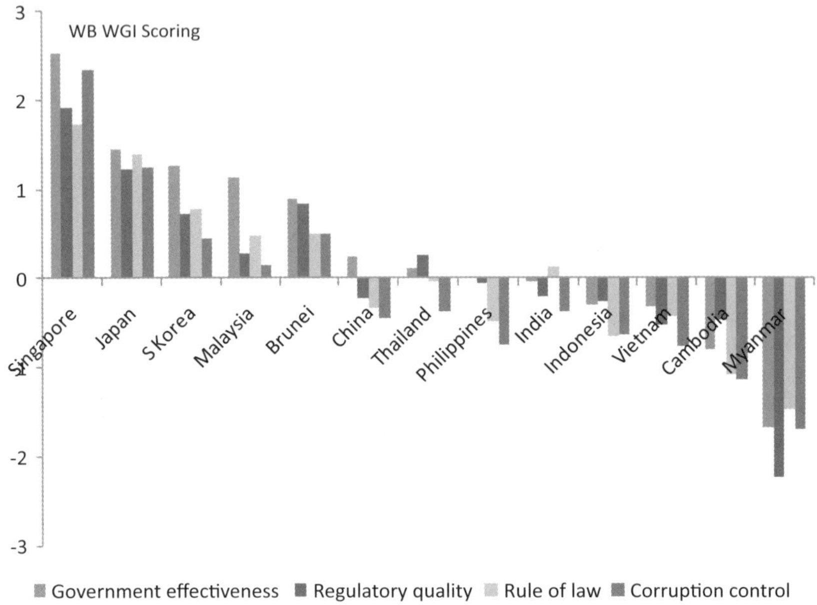

Figure 18 World Bank Worldwide Governance Indicators (WGI) 2008

Source: Kaufmann *et al.*, "Governance Matters VIII: Aggregate and Individual Governance Indicators 1996–2008".

start-ups and licensing, employing workers, property protection, investor protection, trade and capital mobility, fiscal bureaucracy and contract enforcement. The results are generally consistent with the WGI outlined before. However, the report does acknowledge selected reform successes in Indonesia (ease of business start-ups and registering property), the Philippines (ease of getting credit, closing businesses), as well as in Malaysia (improved contract enforcement) and Thailand (business start-ups).

1.5.5 Commercial Strategy and Government Role

One approach that countries in Southeast Asia could adopt is to support more proactively its SOEs and NOCs, as well as their multinational JV partners in commercial strategy development and implementation. The case of the government-linked Economic Development Board (EDB) in Singapore is one example. Starting in the turbulent 1960s, the EDB has helped to attract numerous multinationals to invest in Singapore over the past five decades. In the energy-related field the EDB has helped nurture Singapore's refinery sector and made it among the world's most important refining centres.[71]

Each Southeast Asian country needs to find its own way of helping to attract foreign direct investment and nurture strategic industries. In the context of Southeast Asian countries' energy sector challenges, such a competent and commercially savvy government agency could for example help the respective NOCs and their partners, either evaluate and structure or by evaluating and structuring works I think? optimal government support alternatives including (i) potential seed investment money and fiscal support, (ii) technical, managerial and financial structuring assistance and (iii) advice on transaction structuring and negotiations of long-term contracts with foreign crude suppliers or LNG buyers, engineering, procurement and construction (EPC) contracts, etc. It should be noted that such advice is more relevant for the less commercially experienced SOEs and NOCs. For the more sophisticated ones like PETRONAS and PTT, such government support could also help create market buying power (vis-à-vis counter-parties, i.e. suppliers or buyers), through increased intra-regional coordination and cooperation.

In Indonesia, for example, the request by several multinational gas and LNG project operators for improved fiscal terms and or contract extensions could be addressed by quid pro quo requests for accelerated expansion programmes, stronger involvement of Pertamina and long-term crude supply contracts from the multinational partners' global crude supply portfolio. However, this means that the government decision-makers must be able to work in unison, in the interests of the country, and not be influenced by excessive "rent seeking" from special interest groups.[72]

---

[71] Singapore Economic Development Board (EDB); company website: http://www.sedb.com/edb/sg/.

[72] Again consistent with Robinson's argument that a country's politics and political institutions determine whether or not (industrial) policy has a chance of succeeding. See Robinson, "Industrial Policy and Development: A Political Economy Perspective".

1.6 SUMMARY AND CONCLUSION

The chapter describes a logical, normative approach to energy policy making. Starting with a set of country-specific geographic, demographic, socio-political and economic-industrial determinants, a normative policy grid is derived which helps prioritise the strategies and programmes required to address a country's energy policy opportunities and challenges. The normative approach was examined using the examples of the upstream oil, gas and coal and downstream oil sectors in Indonesia, Thailand and Malaysia, referring to other Southeast Asian countries as selected comparables.

Unsurprisingly, the challenge in Southeast Asian energy policy making is not so much the formulation but the implementation of energy policy. The latter requires synchronisation between industry regulation, trade, investment and fiscal policy, and thus necessitates close coordination across various government ministries and regulatory agencies. Only with concerted efforts geared towards realising a country's economic, security and sustainability objectives can a national energy policy succeed. Given the limitations in financial and institutional resources in selected countries, carefully structured and mutually beneficial arrangements between the government, relevant SOEs and multinational partners are necessary. This requires commercial thinking from the side of the relevant government agencies, which are geared towards the country's overall objectives of industrialisation and development.

CHAPTER 2

EAST ASIA'S ENERGY CHALLENGES: GENERAL ENERGY COOPERATION AND THE QUESTION OF COMPETITION

Christopher Len

2.1 INTRODUCTION

The first East Asia Summit (EAS) was held in Kuala Lumpur, Malaysia, in December 2005. Geographically, this group consists of the 10 Association of Southeast Asian (ASEAN) member countries of Southeast Asia; the three Northeast Asian countries of ASEAN Plus 3, China, Japan and South Korea; as well as Australia, New Zealand and India. It has been described as "a historical event whose future is likely to be as significant as the first Association of Southeast Asian Nations (summit) held in Bali in February 1976".[1]

The aim of the EAS group is to enable dialogue on political, economic and strategic issues for the region. Collectively, this group is meant to

[1]Barry Desker, "Why the East Asian Summit Matters", *Asia Times*, 13 December 2005, http://www.atimes.com/atimes/Southeast_Asia/GL13Ae02.html [Accessed 13 December 2005].

represent a new outward-looking and inclusive regionalism trend within Asia through the membership of all the key Asian economies, with ASEAN at the core. It also covers the important energy importers and exporters within the region and has issued two summit declarations on energy during the second and third East Asia Summits held in the Philippines in January 2007 and Singapore in November 2007 respectively.

This chapter will start by outlining the profiles of the key energy and economic countries within East Asia. It will then demonstrate that analysis of energy developments in East Asia would benefit from Jaewoo Choo's conceptual distinction between general energy cooperation and energy security cooperation. Such a differentiation would help clarify the nature of cooperation over energy issues that are currently underway among East Asian states. The conclusion of this study is that East Asian countries have consistently focused on general energy cooperation, rather than energy security cooperation. Furthermore, it is noted that this emphasis has been the consistent theme for East Asian cooperation over energy issues since 2001, and most recently reflected in the EAS declarations in Cebu and Singapore in 2007.

2.2 ENERGY PROFILES OF KEY COUNTRIES IN EAST ASIA

This section explains the energy profiles of key East Asian countries. The first part focuses on the key energy-importing non-ASEAN countries of the EAS group, namely, China, India, Japan and South Korea. The second part provides a brief overview of the energy profiles of ASEAN member countries, followed by more detailed examinations of the energy profiles of key ASEAN member countries, which have been identified based on the size of their respective populations and economies.

2.2.1 Key Non-ASEAN Member Countries: China, India, Japan and South Korea

Figure 1 shows the growing energy consumption patterns of the non-ASEAN major economies within the East Asia group which are net energy importers, namely China, India, Japan and South Korea. Energy

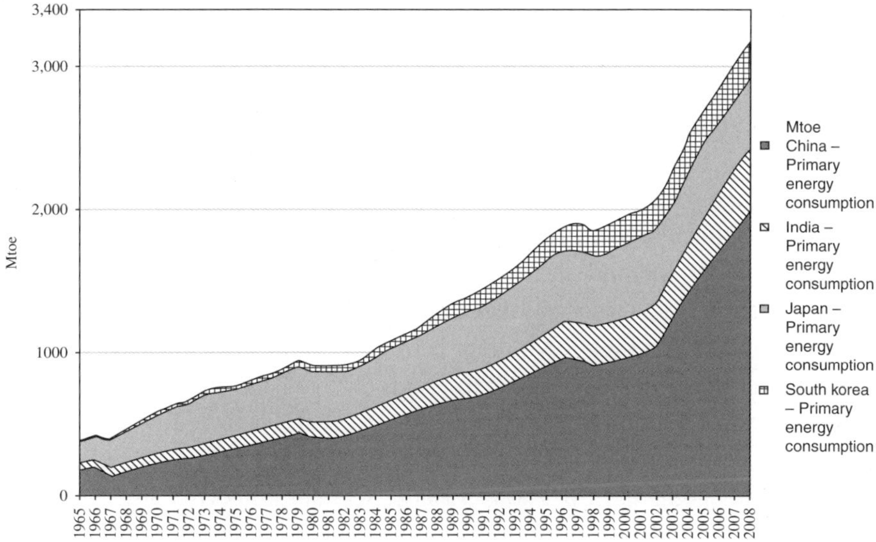

Figure 1 Primary Energy Consumption (Mtoe) of China, India, Japan and South Korea
Source: *BP Statistical Review of World Energy*, 2012.

consumption in these countries has grown rapidly as a result of their economic development and industrialisation over the decades.

China is shown to register the highest primary energy consumption growth rate in Figure 1. It is today one of the world's most dynamic economies and the world's most populous country with nearly one fifth of the world's total population. As of 2009, it is estimated that China ranks third in GDP (purchasing power parity), behind the European Union and the United States.[2] It registered an average growth of 10% between 2000 and 2008 and its economy grew by 8.7% in 2009 despite the global financial crisis caused by the credit crunch.[3]

Coal makes up 70% of China's total primary energy consumption and the country is the world's largest producer and consumer of coal. It started to import coal in 2002 and some projections indicate that it could become

[2]Central Intelligence Agency, *CIA — World Factbook*, 2009, https://www.cia.gov/library/publications/the-world-factbook/rankorder/2001rank.html [Accessed 20 January 2010].

[3]"China Grows by 8.7 Percent in 2009", CNN, 20 January 2010, http://www.cnn.com/2010/BUSINESS/01/20/china.GDP.annual/index.html [Accessed 21 January 2010].

a net coal importer in the years ahead. However, it should be specified that China's coal imports are not driven by the lack of domestic coal reserves, but are largely due to the price spread between local and international coal prices, including international freight rates.[4]

Oil is the second-largest source, accounting for 20% of China's total energy consumption. China grew from being a net importer of oil in 1996 to become the third-largest importer by 2006. It is the world's second-largest consumer of oil behind the United States. The US Energy Information Administration (EIA) forecasts that China's oil demand will continue to grow, with oil demand reaching 8.3 million barrels per day (bbl/d) in 2010, up from 7.8 million bbl/d in 2008. This represents 31% of the projected world oil growth in non-OECD countries from 2008 to 2010. The Middle East remains China's largest source of oil imports. The EIA cited FACTS Global Energy statistics indicating that in 2008, 50% of China's imports came from the Middle East, 30% from Africa. Saudi Arabia and Angola are the two largest sources of oil imports, accounting for over one third of China's total crude oil imports. In order to diversify its import sources, China has actively worked to connect international oil pipelines from neighbouring countries such as Kazakhstan, the Russian Far East and Myanmar to its domestic network.[5]

China is also looking to expand natural gas usage through pipelines and liquefied natural gas (LNG) imports. The Chinese anticipate that the share of natural gas as part of total energy consumption will increase to 10% by 2020. China recently opened a gas pipeline, which connects Turkmenistan through Uzbekistan and Kazakhstan to China's Xinjiang.[6] Another gas

[4] Kate Mackenzie, "Could China Fall out of Love with Coal?", *Financial Times*, 14 January 2010, http://blogs.ft.com/energy-source/2010/01/14/could-china-fall-out-of-love-with-coal/ [Accessed 15 January 2010]; Andrew Peaple, "Will China Light a Fire under Coal Price?", *Wall Street Journal*, 21 August 2009, http://online.wsj.com/article/SB125078103759846543.html [Accessed 15 January 2010]; "China's Coal Imports Predicted to Decline in 2010", *Xinhua*, 3 December 2009, http://www.istockanalyst.com/article/viewiStockNews/articleid/3687142 [Accessed 15 January 2010].

[5] Energy Information Administration (US), *China*, July 2009 version, http://www.eia.doe.gov/emeu/cabs/China/pdf.pdf [Accessed 15 October 2009]; APERC, *APEC Energy Overview 2008* (Tokyo: Asia Pacific Energy Research Centre, 2008), pp. 47–49.

[6] "China Opens Turkmenistan Gas Pipeline", BBC News, 14 December 2009, http://news.bbc.co.uk/1/hi/8411204.stm [Accessed 15 December 2009].

pipeline project to transport gas in Western Siberia into Western China has stalled due to financing and gas pricing issues between Russia's Gazprom and China's China National Petroleum Corporation (CNPC). The country also has a deal to pipe gas from Myanmar. China's imports of LNG started in 2006 and Chinese national oil companies have looked to import LNG from Australia, Indonesia, Malaysia and Qatar through long-term supply contracts and spot cargoes. This puts China in growing competition with the neighbouring Asian economies of Japan, South Korea and Taiwan, which are all heavily reliant on LNG.[7]

India is the second–most populous country in the world, behind China. It is experiencing high economic growth rates like China, posting an average growth rate of 6.6% between the 1988/1999 and 2008/2009 financial years.[8] India is estimated to rank fifth, just behind Japan in global GDP (purchasing power parity) in 2009.[9] Despite a recent slowdown in the economy, India's energy use continues to increase. The country is dependent on coal for half of its total energy consumption, followed by oil (31%), natural gas (8%) and hydroelectric power (6%).[10]

Of its total energy requirements 30% are met through imports, with the country being increasingly dependent on imported oil. In 2007, India was the world's fourth-largest global consumer of oil at 3 million bbl/d in 2008, after the US, China and Japan. India imported 1.8 million bbl/d of oil in 2007, which constituted 68% of its oil consumption. Around 75% of imported oil originated from the Middle East. India was ranked sixth among global net oil importers in 2008. The EIA estimated that India will become the fourth-largest global net importer of oil by 2025, behind the US, China and Japan.[11]

---

[7] *China*, op. cit.; *APEC Energy Overview 2008*, op. cit., pp. 47–49.

[8] Arvind Panagariya, "Building a Modern India", in *Business Standard — India 2010*, edited by T. N. Ninan (New Delhi: Business Standard Books, 2009), p. 1.

[9] *CIA World Factbook*, op. cit., 2009.

[10] Energy Information Administration (US), *India*, March 2009 version, http://www.eia.doe.gov/emeu/cabs/India/Full.html [Accessed 15 October 2009].

[11] *India*, op. cit.; Energy Information Administration (US), *World Regions — Imports: Top World Oil Net Importers, 2008*, http://tonto.eia.doe.gov/country/index.cfm [Accessed 5 December 2009].

Import demand for LNG is also climbing. The power and fertiliser sectors account for nearly 75% of natural gas consumption, and demand for natural gas is expected to grow considerably. India became a net importer of natural gas, imported as LNG, in 2004, and has since relied on both long-term contracts and spot cargoes to secure supplies. In 2006, India was the seventh-largest global importer of LNG, with Qatar being the largest supplier, accounting for nearly 86% of India's LNG imports in 2006. To help meet this growing demand, India has also been considering a number of import schemes via pipelines, namely the Iran-Pakistan-India Pipeline, Turkmenistan-Afghanistan-Pakistan-India Pipeline and pipelines from Myanmar, though all of the projects remain at the proposal stage. The country continues to suffer from electricity shortages. According to the World Bank, roughly 40% of the residences in India are without electricity, while blackouts are common in the main cities.[12]

Both the Chinese and Indian leaderships are faced with the additional challenge of ensuring social stability and this has added to their sense of political urgency to secure adequate supplies from overseas. Both are developing economies faced with rising social expectations from their populations. The two governments are thus under extreme pressure to ensure employment for fear of social unrest erupting, which may spread and disrupt the respective government's overall development efforts. It is therefore especially important to avoid energy supply disruption as this would restrict their economic activities and affect job creation. The two governments have thus become more active in their political and diplomatic roles to secure energy supplies from abroad, with China being particularly expeditious. Furthermore, energy demand in these two countries is also expected to grow significantly since the Chinese and Indian populations have growing aspirations for a higher quality of life. Demand for electrical household items such as television sets, fridges and cars, all of which consume energy and lead to greater energy demand, are expected to grow significantly in the coming years.

Japan is estimated to rank fourth globally in GDP (purchasing power parity), just behind China, in 2009.[13] Japan imports almost all of its crude oil (99%), coal (99%) and natural gas (96%). In 2006, it was the

[12] *India*, op. cit.

[13] *CIA World Factbook*, op. cit., 2009.

third-largest consumer of crude oil, after the US and China, of which 80% was imported from the Middle East. In 2008, it remained the second-largest importer of oil, after the US. It is the world's largest importer of coking coal for steel production, and steam coal for power generation and pulp and paper and cement production. Natural gas is almost entirely met through LNG imports at 76 million tonnes of oil equivalent (Mtoe). This makes it the world's largest importer of LNG, most of which is from Indonesia (23% in 2005), Australia (20%), Malaysia (19%), Qatar (12%) and Brunei (10%). Japan has one of the world's lowest energy intensities among the developed world economies. This is due to the high levels of investment into research and development of energy-efficient technologies at the start of the 1970s. While the East China Sea is believed to contain abundant oil and gas deposits, the exploration and development of such fields have been hampered by territorial disputes with China.[14]

South Korea is estimated to rank 14th in global GDP (purchasing power parity) in 2009.[15] Although oil consumption has remained relatively flat since 2000, South Korea remained the ninth-largest global consumer of oil, and the fifth-largest net importer of oil in 2008, with oil shipped in by tankers, 80% of which was from the Middle East. About a quarter of its oil imports are re-exported as refined petroleum products, mostly to other East Asian countries. Since the introduction of LNG in 1986, the use of natural gas has grown rapidly. In 2006, the country was ranked the world's second-largest LNG importer at 28 Mtoe. Korea is also the world's second-largest importer of both steam and coking coal, after Japan, with imports from China, Australia, Indonesia, Canada, Russia and the US.[16]

Overall, these four major economies share a significant common feature in that they are all highly dependent on imported energy supplies, at

---

[14] Energy Information Administration (US), *Japan*, September 2008 version, http://www.eia.doe.gov/emeu/cabs/Japan/pdf.pdf [Accessed 15 October 2009]; *APEC Energy Overview 2008*, op. cit., pp. 81–83; *World Regions — Imports: Top World Oil Net Importers*, op. cit.

[15] *CIA World Factbook*, op. cit., 2009.

[16] Energy Information Administration (US), *South Korea*, June 2007 version, http://www.eia.doe.gov/emeu/cabs/South_Korea/Full.html [Accessed 15 October 2009]; *APEC Energy Overview 2008*, op. cit., pp. 89–90; *World Regions — Imports: Top World Oil Net Importers*, op. cit.

affordable prices, to keep their economies functioning. As a result of the high level of import dependence, energy has thus become an important national security issue for their governments.

2.2.2 Key ASEAN Member Countries

ASEAN member countries present a rather varied profile due to their diverse political, economic and social development models. The total population of the ASEAN region, which represents about 8.7% of world population, accounts for 3.3% of the world's total primary energy consumption.[17] It has been noted that some of the economies remain pre-industrial, and many households still rely heavily on biomass for cooking and heating. However, commercial consumption is expected to grow over the next two decades because of the general regional trend towards modernisation, industrialisation and urbanisation.[18] The natural energy resource endowment of the member countries also differs vastly, with Singapore having virtually none, while Brunei and Indonesia are rich in such resources. Table 1 provides an overview of the region's energy consumption patterns vis-à-vis their social and economic profiles.

This study will not detail the energy profiles of all 10 ASEAN member countries. Instead, it will highlight the profiles of key ASEAN member countries Thailand, Indonesia, Vietnam, the Philippines and Malaysia. These countries have been selected based on the size of their populations and economies.

Thailand is highly dependent on energy imports though it used to be even more so. Net energy imports went down from 96% in 1980 to 58% in 2006. Its total primary energy supply was 83,863 kilotonnes of oil equivalent (ktoe) in 2006, the top three energy types being oil (57%), gas (26%) and coal (14%). Oil accounted for 62% (34,451 ktoe) of total energy consumed in 2006.[19] The country has implemented a series of

---

[17]Elspeth Thomson, "Southeast Asia's Energy and Security Challenges", in *Energy and Security Cooperation in Asia: Challenges and Prospects*, edited by Christopher Len and Alvin Chew (Nacka-Stockholm: Institute for Security and Development Policy, 2009), p. 22.

[18]Thomson, op. cit. pp. 22–23.

[19]*APEC Energy Overview 2008*, op. cit., pp. 167–168.

Table 1 ASEAN Energy and Economy Indicators in Perspective

	Total Primary Energy Consumption 2006 (Quadrillion BTU)	Population 2007 estimates (Million)	Population Growth 1997–2007 (%)	Per Capita Energy Consumption 2005 (Million BTU)	Per Capita Energy Consumption Growth 1997–2007 (%)	Agricultural Sector, % Total Value Added 2006	Industrial Sector, % Total Value Added 2006	Service Sector, % Total Value Added 2006	Per Capita GNI (USS) 2003
Brunei	0.177	0.37	2.20	482.1	11.48	1.1	67.9	31.0	20,823
Cambodia	0.010	14.00	1.79	0.7	0.43	29.6	29.2	41.2	245
Indonesia	4.149	234.69	1.41	17.9	3.76	13.7	42.4	44.0	599
Laos	0.023	6.52	2.51	3.6	2.77	46.8	27.6	25.7	
Malaysia	2.557	24.84	1.95	104.8	3.49	8.8	50.0	41.2	3,312
Myanmar	0.236	47.37	5.03	5.0	8.44	52.6	13.4	34.0	
Philippines	1.271	91.08	2.00	14.2	3.09	14.2	32.1	53.7	920
Singapore	2.142	4.55	1.80	476.8	5.62	0.1	33.0	66.9	20,066
Thailand	3.741	65.07	0.76	57.9	6.33	10.7	44.6	44.7	1,838
Vietnam	1.404	85.26	1.14	16.6	10.99	21.7	40.2	38.0	392

Source: EIA website http://www.eia.doe.gov/emeu/international/populationan dgdp.html and www.eia.doe.gov/emeu/international/energyconsumption.html (February 27, 2009); Economic and Social Commission for Asia and the Pacific, *Statistical Yearbook for Asia and the Pacific 2007* (New York: United Nations, 2007), p. 88; CIA World Factbook and Nationmaster at www.nationmaster.com/graph/eco_gro_nat_inc_percap-gross-national-income-per-capita (accessed February 27, 2009).

Source: Thomson, op. cit., p. 36.

energy policies aimed at reducing its heavy reliance on energy imports while also ensuring a sound environmental energy strategy.

Measures include acceleration of domestic exploration and development of oil, gas and electricity; enhancement of relations with energy-exporting countries (including its neighbours Myanmar, Cambodia and Indonesia); undertaking a feasibility study of nuclear energy's potential; diversification of fuel types to mitigate risk; promotion of the petrochemical and oleochemical industry for integration with biofuels development in order to generate new industries; promotion of research and development of all forms of alternative energy such as biofuels (gasohol, biodiesel) and natural gas for vehicles (NGV), and renewable energy; support for the Clean Development Mechanism (CDM) to reduce greenhouse emissions; and energy market reform to ensure transparency and fairness and reduce distortions in domestic energy pricing.[20]

Indonesia is rich in natural resources and oil, gas and coal have traditionally played an important role in its economy, particularly as a foreign exchange earner. In 2006, the total primary energy supply of Indonesia was 130,520 ktoe, comprised mainly of oil (44.7%), natural gas (29.4%) and coal (20.9%).[21] It became a member of the Organization of the Petroleum Exporting Countries (OPEC) in 1962, but withdrew in 2008 due to the decline in its production of oil. In 2007, Indonesia enacted its first legislation on energy, designated the Energy Law. The law made clear the government's principles "with regard to utilisation of energy resources and final energy use, security of supply, energy conservation, protection of the environment with regard to energy use, pricing of energy, and international cooperation".[22] Prior to the Energy Law, the government drafted energy policies (National Energy Policy) through public consultation before legislation and promulgation. The National Energy Policy detailed the country's energy policy in its Blueprint for National Energy Management 2005–2025, which was issued in July 2005. The document focused on ensuring supply of energy through domestic supply, optimising energy production and implementing energy conservation; energy efficiency and

[20] Ibid., pp. 168–175.

[21] Ibid., p. 65.

[22] Ibid., pp. 67–80.

diversification; pricing energy at economic levels with due regard to the needy members of society; and protection for the environment. The country's basic strategy is to encourage self-reliance by encouraging domestic exploration and exploitation of oil, gas, coal and geothermal sources.[23]

The government has also implemented energy conservation measures with the National Master Plan for Energy Conservation enacted in 1995. The plan included programmes for training, efficiency awards, energy management and industrial audits. It also provided tax reductions and soft loans as incentives. In 2000, it announced a target to reduce energy intensity by 1% per year. Indonesia, along with all other ASEAN member countries, except Brunei, adopted a voluntary building energy code initiated in 1992.[24]

In recent years, the country has also been looking into biofuels, renewable energy, energy efficiency and nuclear power. In 2006, the President mandated (Presidential Decree No. 10/2006) the formulation of a blueprint on biofuel development and in 2008 issued a regulation (Ministerial Regulation No. 32/2008) regarding the supply, use and commerce of alternative biofuels as an alternative energy. Special Biofuel Zones (SBZs) have been designated, covering at least 10,000 hectares in Java and 100,000 hectares outside Java, for the cultivation of crops to be used for biofuel. It has also taken measures to invest in renewable energy, particularly solar, and expanded efforts in recent years to improve energy efficiency (Presidential Instruction No. 2/2008). In 2007, the government prepared a Nuclear Power Development Preparatory Team to prepare for and work towards the country's aim to build nuclear power plants.[25]

Vietnam began market reforms in 1986 and this led to rapid economic development. Energy is a crucial factor in Vietnam's economic development plan since its growth is increasingly reliant on its industrial base. Its total primary energy supply in 2006 was 43,628 ktoe, which was an increase of 7.05% from the previous year. Its energy mix consists of oil (29.3%), coal (20.7%), natural gas (11.8%) and other sources (34.1%).

---

[23] Ibid., pp. 71–78.

[24] Ibid., pp. 70–71.

[25] Ibid., pp. 78–79.

The country is rich with diverse fossil energy resources such as oil, gas and coal, as well as renewable energy such as hydro, biomass, solar and geothermal. The country became a net exporter of energy in 1990 with crude oil and coal being the main exports. Vietnam also has gas reserves that are said to be more promising than its oil reserves. However, its gas potential has been complicated by its protracted offshore territorial disputes in the South China Sea with China.[26]

The country is expected to transform from being a net exporter of energy to a net importer beyond the year 2010 as consumption is expected to rise by 4.3 times in the coming two decades.[27] As a result, the government has been preparing a long-term policy to ensure its supply of energy. This consists of implementing energy efficiency and conservation initiatives, increasing the country's ability to carry out domestic exploration so as to reduce reliance on imports, supporting Vietnam's national oil company in expanding its oil and gas activities overseas, and developing clean fuels, particularly from nuclear and renewable sources (wind, solar and hydropower).[28]

The Philippines' total primary energy supply, excluding traditional biomass fuels, was 31,126 ktoe, of which 54.7% (17,019 ktoe) was imported. In 2006, its energy sources were oil (68.1%), others (23.6%), coal (8%) and gas (0.3%). The 2006 Philippine Energy Plan indicates that from 2007 to 2014, the economy will grow by 3.3% annually, with petroleum making up the bulk of the final energy demand with an average share of 39%. This will be closely followed by biomass with a 38% share.[29] In order to become less reliant on energy imports, the government has targeted a 60% energy self-sufficiency level by 2010, which is to be maintained through 2014. It is thus increasing indigenous oil and gas exploration and coal production. It is also looking to increase renewable energy capacity (biomass, wind and solar), together with the use of alternative fuels (biofuels) and research into nuclear energy.

---

[26] Ibid., pp. 188–191.

[27] "Vietnam to be Net Importer by 2015", VietNamNet Bridge, 24 January 2008, http://english.vietnamnet.vn/biz/2008/01/765806/ [Accessed 5 December 2009].

[28] *APEC Energy Overview* 2008, op. cit., pp. 192–200.

[29] Ibid., pp. 131–132.

Energy efficiency and conservation programmes will also be further developed.[30]

Malaysia's total energy supply in 2006 was 59,318 ktoe with oil (44%), gas (44%) and coal (10%) being the main fuel types. While it remains a net energy exporter, it is aware that its oil and gas reserves will be depleted soon and is therefore looking at energy-efficient initiatives and the development of viable alternative energy sources, such as solar, wind, biofuels and nuclear.[31]

Another country worth mentioning is Myanmar. While it has notable oil and gas deposits, international investors and neighbouring governments generally avoided dealing with the country because of the military regime in place. With the election of President Thein Sein and reforms taking place since March 2011, this could well change.

2.3 DISTINGUISHING GENERAL ENERGY COOPERATION FROM ENERGY SECURITY COOPERATION

The importance of recognising the different types of cooperation being undertaken in Asia was first pointed out by Jaewoo Choo in his study on Northeast Asian energy cooperation. He did so by highlighting the conceptual distinction between general energy cooperation (GEC) and energy security cooperation (ESC).[32] In his study, Choo distinguished the former as being broader and more technical in scope, while the latter concept more constricted and politically sensitive. Thus, GEC may include energy security issues, but energy security issues would not cover interest in other energy issues, such as renewable energy, energy technology and the environment.[33]

He noted that governments have tended to refer to general energy cooperation and energy security cooperation interchangeably when referring to their long-term energy policy statements. As a result, "improperly

[30] Ibid., pp. 132–140.

[31] Ibid., pp. 97–100.

[32] Jaewoo Choo, "Energy Cooperation Problems in Northeast Asia: Unfolding the Reality", *East Asia*, 23, no. 3 (2006), pp. 91–106.

[33] Ibid., p. 92.

understood and ill-defined concepts (ideas) inevitably and easily result in misconceived policy, orientation, and implementation". This makes the search for common ground difficult due to misunderstandings on what each party is prepared to commit to, leading to the impression of cooperation as "unthinkable, if not impossible, among these states".[34]

By distinguishing the two concepts, he argued that a clearer picture emerges as to the types of cooperation governments would be willing to engage in, with whom and why. Choo defined energy security as "the maintenance of sufficient energy supplies, prices commensurate with purchasing power, and guaranteed safe delivery of energy resources".[35] He pointed out that the concept does not merely refer to economic interests. Rather, it has been securitised at the national level and is today part of a state's national security interests, particularly with respect to Asian governments. As a result, "economic value has lost ground in foreign-policy making communities of the regional states".

Some Asian countries thus regard energy security interests in zero-sum calculations due to the existing inter-state political rivalry and distrust within the region.[36] From this, one could deduce that ESC is essentially about the allocation of energy supplies between different state actors. While it would make sense to governments of importing countries to engage in such cooperative arrangements with suppliers, they are less willing to do so with other importing countries because of their zero-sum mentality.

Choo's idea of GEC meanwhile is drawn from Michael Jefferson's chapter on energy policies for sustainable development in a UNDP report published in 2000.[37] Energy cooperation within Jefferson's context is "more concerned with the provisions of energy services through improved management of international energy markets, wider availability and higher quality of energy resources and the expansion of choices in order

---

[34] Ibid., pp. 92–93.

[35] Ibid., p. 93. Choo drew this definition from Philip Andrews-Speed, "Energy Security in East Asia: A European View", paper presented at the Symposium on Pacific Energy Cooperation, Tokyo, 12–13 February 2003.

[36] Choo, op. cit., p. 95.

[37] Michael Jefferson, "Energy Policies for Sustainable Development", in *World Energy Assessment: Energy and the Challenge of Sustainability*, edited by José Goldemberg (New York, NY: United Nations Development Programme, 2000), pp. 415–452.

Table 2 The Need for a New Energy Paradigm

Traditional Paradigm	Emerging Paradigm
Energy considered primarily as a sectoral issue	Greater consideration of social, economic, and environmental impacts of energy use
Limitations on fossil fuels	Limitations on the assimilative capacity of the earth and its atmosphere
Emphasis on expanding supplies of fossil fuels	Emphasis on developing a wider portfolio of energy resources, and on cleaner energy technologies
External social and environmental costs of energy use largely ignored	Finding ways to address the negative externalities associated with energy use
Economic growth accorded highest priority (even in prosperous economies)	Understanding of the links between economy and ecology, and the cost-effectiveness of addressing environment impacts early on
Tendency to focus on local pollution	Recognition of the need to address environmental impacts of all kinds and at all scales (local to global)
Emphasis on increasing energy supply	Emphasis on expanding energy services, widening access, and increasing efficiency
Concern with ourselves and our present needs	Recognition of our common future and of the welfare of future generations

Source: Michael Jefferson, "Energy Policies for Sustainable Development", in *World Energy Assessment*: *Energy and the Challenge of Sustainability*: http://bbs.cenet.org.cn/uploadimages/20053810181399130.pdf, p. 418.

to achieve sustainable growth".[38] It therefore has an expanded scope of energy policy goals, which incorporates sustainable growth imperatives at the local, regional and global levels. Table 2 summarises Jefferson's changing energy paradigm, and shows how energy policy concerns have been expanding to include sustainable development considerations.

Such an expanded conceptualisation of interests with the new and expanded focus facilitates cooperation dynamics. For instance, it has provided a new rationale for countries with the know-how to help other

[38]Choo, op. cit., p. 93.

countries, particularly those within the same region, to upgrade their oil refinery technologies to reduce inefficiencies and pollution. Such technical cooperation does not undermine one's own fundamental energy security interests, but is aimed at addressing transnational energy-related problems.

GEC is thus suitable for addressing common energy-related problems through synergies, for example joint research in developing alternative fuels, or in drawing financial capital for projects because of the enlarged market size. It is believed that such a process could provide momentum for greater socialisation, further expanding the scope of energy cooperation, and potentially — though not always the case — lead towards energy security cooperation between the parties.[39] Understood as such, GEC could play a functionalist role, providing the first practical steps which could help facilitate ESC; although whether ESC does take place would depend on a host of other issues, which could propel or obstruct such cooperation.

2.4 EAST ASIA'S FOCUS ON GENERAL ENERGY COOPERATION

In this section, the author will demonstrate that the cooperation framework currently underway within the EAS grouping is in fact general energy cooperation, not energy security cooperation. The purpose is not to analyse the viability or the follow-up measures undertaken by the respective governments following the Cebu and Singapore declarations. Instead, it is to clarify the nature of cooperation that is currently underway, and to enable us to make better sense of the political intentions and priorities of cooperation among East Asian countries.

Energy has been cited as an avenue for cooperation within the East Asian context from as early as 2001. The East Asian Vision Group (EAVG), set up in 1998, and composed of members from ASEAN Plus 3, submitted a vision report, "Towards an East Asian Community", to the ASEAN Plus 3 leaders in October 2001.[40] In its key recommendations, it

---

[39] Ibid., p. 94.

[40] *Towards an East Asian Community: Region of Peace, Prosperity and Progress*, East Asian Vision Group Report, 2001, http://www.mofa.go.jp/region/asia-paci/report2001.pdf [Accessed 1 December 2009].

stated that one dimension of environmental cooperation should include "… joint development and exploration of new sources and supplies of energy within the region and embarking on efficient use of energy".[41] The main report was very much focused on cooperation to develop cleaner forms of energy. This was noted in paragraph 70:

> The Vision Group proposes an environmentally friendly and sustainable East Asia. It calls upon leaders to strengthen and increase efforts towards institutionalizing environmental and energy cooperation in the region.

Paragraph 86 further advocated that

> Research and development should be stimulated for the exploration of alternative sources of cleaner energy, including renewable energy generation. Given the drawbacks of fossil fuels, East Asian countries should cooperate for the development of solar, wind, hydroelectric and nuclear fusion energy. There should be more cooperative regional efforts to establish renewable energy generation systems.

In January 2002, Japanese Prime Minister Junichiro Koizumi announced Tokyo's intention to form an East Asian Energy Community, using the ASEAN Plus 3 framework. This was meant to enhance Japan's leadership role with ASEAN on non-traditional security issues within East Asia.[42]

In November 2002, the East Asia Study Group (EASG), set up in 2000, submitted a document during the ASEAN Plus 3 summit, titled, "Final Report of the East Asian Study Group".[43] Besides emphasising the ASEAN energy networks, the EASG authors reiterated the conclusions of the earlier EAVG group, and framed East Asia's proposed energy

---

[41] *Towards an East Asian Community*, op. cit., p. 4.

[42] Gaye Christoffersen, "East Asian Energy Cooperation: China's Expanding Role", *China and Eurasia Forum Quarterly*, 6, no. 3 (2008), pp. 150–151.

[43] *Final Report of the East Asian Study Group, ASEAN + 3 Summit*, 4 November 2002, Phnom Penh, Cambodia, http://www.aseansec.org/pdf/east_asia_vision.pdf [Accessed 10 July 2007].

strategy within the environmental context. Paragraph 37 of the Final Report stated that

> The EASG attaches importance to energy security for the sustainable development of the East Asian economy through cooperation among East Asian countries. Although there is no plan as yet for cooperation on nuclear energy, comprehensive cooperation through a framework policy will greatly enhance cooperation among East Asian countries in consideration of their existing national policies. Cooperation in research and development of alternative and/or cleaner energy sources can be included in the comprehensive strategy and policy on energy development in the East Asian region.

In 1998, ASEAN adopted the Hanoi Plan of Action as part of its ASEAN Vision 2020 target which was introduced in 1997. This included a strategic plan of action for energy cooperation known as the ASEAN Plan of Action for Energy Cooperation (APAEC) 1999–2004 in order to facilitate energy cooperation among ASEAN members. In 2004, ASEAN issued a follow-up APAEC 2004–2009.[44] APAEC 2004–2009 spoke of an "evolving plan of action", making reference to the ASEAN Vision 2020 which called for, among other things, first, the establishment of the ASEAN Power Grid and Trans-ASEAN Gas Pipeline; second, promotion of cooperation in energy efficiency and conservation, as well as development of new and renewable energy resources. It also envisioned "a clean and green ASEAN with fully established mechanisms for sustainable development to ensure protection of the environment, the sustainability of its natural resources, and the high quality of life of its peoples".[45] The document went on to frame a new plan of action "in the context of sustainable development" as echoed during the UN World Summit on Sustainable Development in Johannesburg, South Africa, in 2002. Thus, the document, in reflecting the sentiments from the World Summit, considered probable mid-term scenarios in the

---

[44] *ASEAN Plan of Action for Energy Cooperation (APAEC), 2004–2009*, http://www.aseansec.org/pdf/APAEC0409.pdf [Accessed 5 December 2009].

[45] Ibid., p. 3.

ASEAN region, based on the combination of "economic, social and environmental" considerations.[46]

Significantly, APAEC 2004–2009 indicated ASEAN's intention to tap external bilateral and multilateral partners for its regional energy initiatives, namely the European Union, Japan, the ASEAN Plus 3 parties, Australia and the Energy Charter Secretariat. It was affirmed that the programme areas of APAEC 1999–2004 would be maintained, namely (1) the ASEAN Power Grid, (2) the Trans-ASEAN Gas Pipeline, (3) institutionalisation, improvement and promotion of coal usage, (4) energy efficiency and conservation, (5) renewable energy, while a new (the sixth) programme called Regional Energy Policy and Planning (REPP) was also incorporated.

Despite the fact that the US was not identified as a partner in the APAEC 2004–2009 document, ASEAN nonetheless approached it to be a partner for its regional energy initiative. During the 19th ASEAN-US Dialogue held in May 2006, ASEAN urged the US to promote cooperation in the field of energy efficiency, and in affordable technologies for renewable and alternative energy. This led to a five-year action plan in July 2006 with energy cooperation listed as one area of collaboration.[47]

In January 2007, the Cebu Declaration on East Asian Energy Security was announced at the second EAS.[48] It emphasised energy policies based on sustainable development and presented a broad outline of intended measures focused on the following: the promotion of alternative clean and renewable energy, reduction of emissions through use of technology, improvement of energy efficiency and conservation, use of clean coal and clean coal technologies, strategic fuel stockpiling, engagement in both bilateral and regional research and development, and exploring means for financing such projects.

According to one author, this declaration "underscores the fact that the search for alternative fuel sources has been prompted by an awareness of the diminishing supply of fossil fuels, the unstable global prices of oil and

---

[46] Ibid., p. 4.

[47] Christoffersen, op. cit., p. 151.

[48] Cebu Declaration on East Asian Energy Security, Cebu, The Philippines, 15 January 2007, http://www.aseansec.org/19319.htm [Accessed 9 December 2009].

the worsening problems of the environment".[49] The Cebu Declaration has been criticised for being a "zombie initiative" because only a broad set of intended measures were announced without any concrete mechanisms being set.[50] Nonetheless, it reinforced the ideational theme and intention of the East Asian leaders on the form of energy cooperation to be carried out within the East Asian framework, which was consistent with the proposals set forth by the EAVG and EASG.

The emphasis on GEC was again evident at the third EAS held in Singapore in November 2007, with attention focused on the threat of climate change. The Singapore Declaration on Climate Change, Energy and the Environment acknowledged climate change as a shared concern for EAS members and stressed "the important role that the EAS can play in carrying out collective action to address these challenges for mutual benefit and the common good".[51] The Singapore Declaration focused on measures that would mitigate the environmental costs of energy consumption, promote sustainable environmental initiatives, and better manage the risks of environmental disasters by building contingencies. It even set the goal of implementing measures "recommended by the EAS Energy Ministers, including the formulation of voluntary energy efficiency goals by 2009". It also tasked the relevant member countries to follow up on the summit declaration at the inaugural EAS Energy Ministers' Meeting held in Singapore in August 2007.

What then of energy security cooperation with the EAS group? Some may point to the Trans-ASEAN Gas Pipeline Infrastructure Project (TAGP) as an example of ESC at work within ASEAN. The TAGP plan is to construct an integrated natural gas pipeline network to connect gas reserves in the Gulf of Thailand, Myanmar and Indonesia to the rest of the ASEAN region. This is supposed to be a multi-billion-dollar project to build a series of natural gas pipelines spanning the 10 ASEAN countries

---

[49] Renato Cruz de Castro, "Assessing the Cebu Declaration on East Asian Security: Issues and Challenges in Regional Energy Cooperation", in *Energy and Security Cooperation in Asia: Challenges and Prospects*, edited by Christopher Len and Alvin Chew (Nacka-Stockholm: Institute for Security and Development Policy, 2009), p. 271.

[50] Christoffersen, op. cit., p. 152.

[51] Singapore Declaration on Climate Change, Energy and the Environment, Singapore, 21 November 2007, http://www.aseansec.org/21116.htm [Accessed 9 December 2009].

to connect a population of 677 million people spread across 4.5 million km^2 of land. The idea was first discussed by individual member countries in 1986, with formal plans announced in 1990. However, feasibility plans only started in 1998, and it was only in 2002 that ASEAN signed a memorandum of understanding binding ASEAN countries to construct the TAGP.

Southeast Asian countries have to date invested US$14 billion to construct 4,000 km of pipelines. Some advocates have even argued that the network could connect gas markets in China, Japan and India to create the largest gas network in the world, constituting US$93.6 billion worth of investments.[52]

However, Sovacool has pointed out that the TAGP has encountered a wide range of problems in its implementation due to technical, economic, legal, social and environmental reasons.[53] One significant observation he made was that ASEAN countries are becoming more protectionist and nationalist in their attitude towards energy resources. The notable trend has been to shift towards domestic control and usage of energy resources, instead of exporting them for trade.[54] He described the TAGP as "less of a concrete idea and more an evolving concept" and noted that "if there is a single concise, and overarching conclusion about the TAGP, it is that the project remains still in its formative stages, despite the past two decades".[55]

2.5 CONCLUDING THOUGHTS: ENERGY SECURITY COOPERATION AND COMPETITION IN EAST ASIA

There is no denying that a multilateral approach is needed in response to today's energy security challenges. However, what this study has demonstrated is that the East Asian countries have consistently focused

---

[52] Benjamin K. Sovacool, "Energy Policy and Cooperation in Southeast Asia: The History, Challenges, and Implications of the Trans-ASEAN Gas Pipeline (TAGP) Network", *Energy Policy*, 37 (2009), pp. 2356–2358.

[53] Ibid., pp. 2362–2365.

[54] Ibid., p. 2365.

[55] Ibid., p. 2364.

on general energy cooperation as a means to address their energy and environmental concerns. Cooperation in this area seems more palatable since it often focuses on technical and scientific issues, with less strategic and political implications.

Energy security cooperation, on the other hand, is, and will continue to be, more exclusive and limited since states, particularly those in Asia, tend to think of cooperation within this sphere as having greater strategic implications for national security. This is primarily associated with the acquisition and transport of natural energy resources. Cooperation will largely remain bilateral, between importer and exporter.

The basic problem for the EAS members is the general sense of distrust that continues to pervade among member states, and this is set to hamper multilateral energy security cooperation, particularly among the energy-importing members, in the years ahead. Energy security cooperation will only develop if there are (1) strong and reciprocal political personalities driving the agenda; (2) economic viability and incentives to justify such projects; (3) strong investor confidence to fund such projects; and (4) social legitimacy in ensuring that the projects can proceed without disruption by the local populations.

Instead of ending the chapter sounding fatalistic, the author would like to point out that given the current conditions in East Asia, it should be acknowledged that while competition is currently inevitable, not all forms of competition are bad. In fact, this author argues that competition is an inherent feature within all societies, including international society. Therefore, state leaders should be nuanced in their thinking and be careful not to mistake their role as being to "eradicate competition".

Competition has a functioning role within economic activity as a means to improve efficiency, innovation and choice, all of which are ingredients that facilitate progress. From social and international relations perspectives, it also enables actors to learn to interact and relate to one another, and help decision-makers set and prioritise their goals. The act of competition in fact allows actors to come into contact with, and develop, social rules and international norms as a means of engagement.

The objective should therefore not be to end energy-related competition, but to address zero-sum thinking in such engagements. In this aspect, political leaders should focus more on laying down the rules of regional engagement, strengthen preventive diplomacy and confidence-building mechanisms and be prepared with coordinated responses for conflict management in the event of crisis situations. This is to ensure that competition over access to energy resources does not spiral into political brinkmanship or military conflict, leading to a destabilised region.

CHAPTER 3

CONTEMPORARY JAPANESE-RUSSIAN ENERGY COOPERATION: PROBLEMS, CURRENT DEVELOPMENTS AND PERSPECTIVES

Svetlana Vassiliouk

3.1 INTRODUCTION

Given the geographical proximity, complementary economic needs, and desire to diversify their respective energy policies, Japan (one of the world's top energy consumers) and Russia (one of the world's top energy producers) should naturally seek to expand their energy relations with each other. The development of Russia's rich energy base and infrastructure in the remote areas of Eastern Siberia and the Far East to open up new export markets in Asia has become an important direction for Russia's overall economic policy as stipulated in the Eastern Gas Program and Energy Strategy to the year 2030. This strategically critical, yet enormous task requires a long-term financial commitment and extensive technological investment, which is difficult for the Russian government to tackle on its own. Furthermore, given the deteriorating demographic situation and growing Russian apprehension toward Chinese expansion in this area, Japan could become a natural partner in the development of these regions. Russia's convenient location and natural resource endowment could greatly benefit Japan, which is focusing on securing reliable alternative energy supplies in close proximity. Moreover, as both Japan's and Russia's national energy strategies stress similar goals, such as the improvement of energy efficiency, promotion of renewable energy resources, conservation of energy resources, and advancement of clean technologies to facilitate emission reductions, the two countries are now presented with new opportunities to expand their bilateral energy cooperation in these areas, in addition to traditional trade in oil and natural gas.

It is important to examine the historical precedents of bilateral cooperation between the two countries, focusing on major joint projects, as the lessons learned from these energy cooperation initiatives could provide important insights for the realisation of ongoing and future joint projects. However, the assessment of their energy cooperation should also include a brief analysis of their political and diplomatic bilateral relations as well as the developments in their respective relations with other important players in the international arena (particularly China and the United States), since all of these factors have important implications for the realisation of their joint initiatives in the energy field.

Contemporary Japanese-Russian relations, including bilateral energy cooperation, have undoubtedly had a very complex and contentious history and have been affected, to a certain degree, by the developments in their political relations. Thus, while the two countries' trade and economic ties have recently been enjoying robust growth,[1] they are yet to reach their full potential as each country's share in the other's total trade turnover is still very low. Moreover, Japanese-Russian political relations remain hostage to the unresolved dispute over the four Kuril Islands off Hokkaido as a notorious legacy from the WWII period. Despite Japan and the Soviet Union signing a joint declaration and de facto normalising their diplomatic relations in 1956, the two countries, having failed to conclude their peace agreement and final border settlement, are still at odds and yet to fully normalise their relations with each other.

[1]Despite the setbacks caused by the financial crisis in late 2008, trade between Japan and Russia has been steadily growing, reaching the $30 billion mark by the end of 2009 (Bloomberg, 17 February 2009). However, in 2009, in terms of exports to and imports from Japan, Russia's rankings were 27th and 17th respectively; whereas Japan has become one of Russia's top largest Asian trade partners (JETRO, 2009).

Based on the close linkage between politics and economics in the two countries' contemporary energy policies, it is plausible to argue that through expansion of their energy cooperation, Japan and Russia should be able to improve their political relations. Unlike in the Cold War era, unresolved political problems do not seem presently to have a significant effect on Japan-Russia economic and trade cooperation. These political hurdles are not the only obstacles that the two countries must overcome to reach full potential in their bilateral economic relations, including energy cooperation. It is crucial to continue expanding bilateral relations at various levels, thereby creating opportunities for both countries to deepen their dialogue in search of timely, effective, and mutually acceptable solutions to their outstanding problems, including the Kuril territorial dispute. Furthermore, building their relations based on mutual trust and removing the remaining obstacles to better understanding would help improve bilateral relations in all other spheres as well, including deeper economic integration and energy cooperation.

3.2 HISTORY OF BILATERAL COOPERATION: LESSONS AND PROBLEMS FROM THE PAST

In spite of various political, strategic, and economic concerns that were especially evident during the Cold War period, Japan and the former Soviet Union had begun discussions about the possibility for joint development of energy resources, particularly in the Soviet Far East and Western Siberia, in the late 1960s.[2] In addition to the development of crude oil and gas in the region, early bilateral initiatives focused on development projects in transport, infrastructure, and the mining industry, especially coal.

The first Soviet-Japanese joint energy project proposals materialised in the early 1970s. The proposals focused mainly on joint development of energy resources in Western Siberia (for example the Yakutiya Natural

[2] The author of this presentation has written an extensive account of early joint initiatives for energy cooperation in her doctoral dissertation "Energy Politics in Japanese-Soviet/Russian Relations: From Cooperation Initiatives in the 1970s to Cautious Engagement in 1990s" (Hosei University, Tokyo, March 2006) and in related articles, such as "Energy Politics in Japanese-Soviet Relations in the 1970s: Complementarity and Conflict", *Russian and Eastern European Studies*, vol. 35 (March 2007).

Gas and Tyumen Oil Development projects) and the Soviet Far East (for example the Sakhalin Continental Shelf Oil and Gas Exploration Project).

However, with the exception of the Sakhalin oil and gas project, the realisation of these projects failed due to a number of factors. In particular, political tensions and strategic considerations incited by the Cold War rivalry between the Soviet Union and the United States played the most critical role in derailing bilateral energy initiatives and projects (Table 1).

Although Japan was interested in expanding its energy collaboration with the neighbouring USSR, as a US military ally, it also had to consider, and often side with, the US stance on its foreign and economic policy vis-à-vis the Soviet Union. These considerations, exacerbated by the unresolved Kuril Islands dispute (Figure 1), resulted in the shift of Tokyo's policy toward politicisation of its trade and energy cooperation with Moscow, which by the early 1980s evolved into the so-called policy of *seikei fukabun* (inseparability of politics and economics).

Another external political factor that cast a shadow on Japanese-Soviet energy cooperation in the 1970s was the US and Japan's political rapprochement with China in 1972, which the Soviet Union viewed as a direct threat to its strategic interests in East Asia. Not only did China support Japan's territorial claim in the Kuril dispute, but it (along with the US) put pressure on Japan to withdraw from the energy projects in Eastern Siberia (particularly the Tyumen Oil Development Project), arguing that such collaboration would grant the Soviets strategic advantage in the region.

Japan saw US participation as its "insurance policy" to share large-scale credit and investment risks and also sought US political support for its territorial claims in its relations with the Soviet Union. The failure to secure the backing of the United States tapered Japanese enthusiasm for participation in joint energy initiatives with the Soviet Union. Finally, after the 1979 Soviet invasion of Afghanistan, the United States and its allies launched the ideologically driven "economic Cold War" against the USSR. The US immediately reacted by imposing a trade embargo on the USSR (subsequently adopted by the Japanese government), which, in addition to the technological embargo introduced in 1978, further undermined the feasibility of Japanese-Soviet energy cooperation.

In addition to the aforementioned political and ideological factors, numerous problems with project financing due to difficult loan and credit negotiations between Japan and the USSR, a lack of sufficient technology

Table 1 Japan-USSR Joint Energy Initiatives in the 1970s

	Yakutiya Natural Gas Project	Tyumen Oil Development Project	Sakhalin Continental Shelf Oil and Gas Exploration Project
Initiation date	Joint agreement signed in 1974; expected start date: 1982	Initially discussed in February 1972; expected start date: 1981	Joint agreement officially signed in January 1975
Primary project signatories and developers	**USSR**: Ministry of Foreign Trade **Japan**: Tokyo Gas, Natural Gas Kondankai **US**: El Paso Natural Gas and Occidental Petroleum	**USSR**: Ministry of Foreign Trade **Japan**: Petroleum Committee (US declined to participate)	**USSR**: Ministry of Foreign Trade **Japan**: SODECO (created in October 1974 especially for this project; currently a member of Sakhalin-1 consortium) and JNOC **US**: US Gulf Oil as a minor shareholder
Net total investment	Exploration stage investment: Japan and US to invest US$25 billion each	Japan was expected to provide bank loans of US$1.3–1.7 billion	US$237.5 million dollars, including US$150 million in commercial credit for the first 5 years

(Continued)

Table 1 *(Continued)*

	Yakutiya Natural Gas Project	Tyumen Oil Development Project	Sakhalin Continental Shelf Oil and Gas Exploration Project
Project description	25-year term; joint development of Yakutiya gas reserves and related infrastructure; construction of gas pipeline to port Olga on the Sea of Japan and a related LNG plant; Primary target customers are Japan and the US West Coast (LNG).	Japan was expected to supply all necessary equipment for oil exploration, drilling, and delivery of crude and to construct a seaport facility for crude shipments from Russia; in exchange, Japan would be compensated with 25 million tonnes of crude annually for 25 years	One of the first compensation-based projects; initial term was set at 10 years; Japan was expected to provide capital investment and commercial credit to finance Soviet drilling and exploration activities in Sakhalin; in compensation, the Soviet side would provide Japan with 50% of the obtained oil during the loan term and for another 10 years after its expiration
Status/ Outcome	Due to various (mostly political) considerations, the trilateral joint venture was officially dismantled in 1980	Excessive project costs, especially for infrastructure development and great distances, made this project prohibitive; strategic and political reasons (protests by the US and China regarding the BAM railroad construction to deliver crude oil to the Pacific coast) also prevented this project from materialising and led to its abandonment by the mid-1970s	Due to political, economic, and technical considerations (such as lack of necessary drilling equipment and prohibitive oil extraction costs), the project had stumbled by the late 1970s (especially after the Soviet invasion of Afghanistan and subsequent US trade embargo policies); it was not completely abandoned and instead re-emerged during the mid-1990s as the Sakhalin-1 project

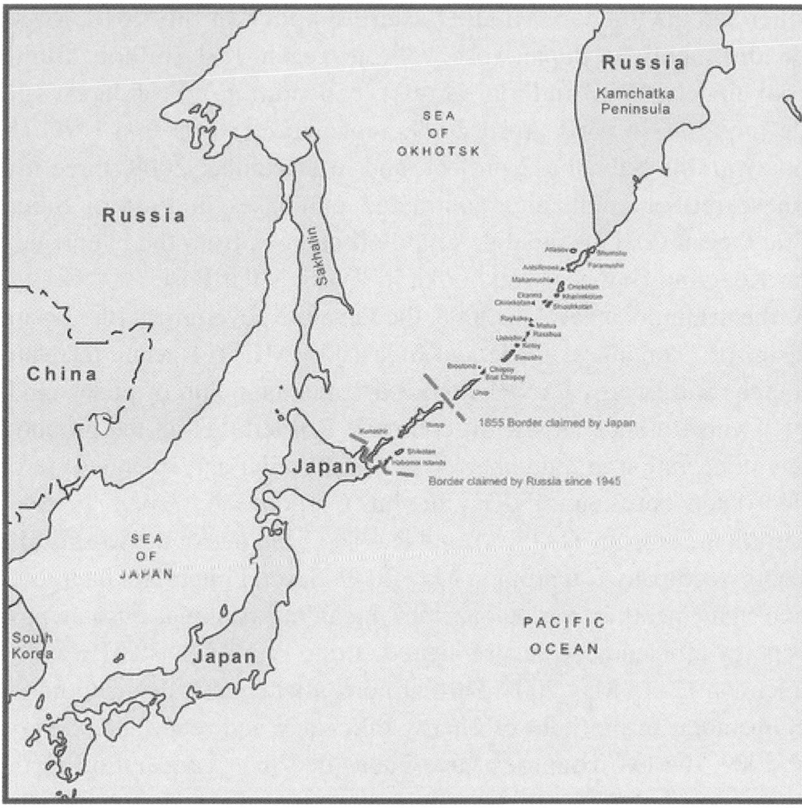

Figure 1 Map of the Disputed Territories between Japan and Russia

Source: The Reviewer Library. <http://sadowa.tumblr.com/post/30701826673/a-map-of-the-kuril-islands-border-dispute-between>. Accessed on December 14, 2010.

and equipment, poorly developed infrastructure, and difficult climate conditions in Eastern Siberia and the Far East made it very difficult for Japanese-Soviet energy projects to materialise.

3.3 OVERVIEW OF CURRENT DEVELOPMENTS IN JAPANESE-RUSSIAN ENERGY COOPERATION

Japanese-Russian energy cooperation has been gradually improving since the materialisation of the Sakhalin-1, Sakhalin-2, and related oil and gas projects in the Russian Far East in the mid-1990s. Recently, Russia has overtaken Iran to become Japan's fourth-largest crude oil supplier and the top non–Middle Eastern supplier. In July 2010, Russia's crude oil supply to Japan rose 93% to reach 1.61 million kilolitres (approximately 10.13 million barrels), constituting 9% of Japan's total crude imports.[3] In early April 2009, Japan received its first LNG shipments from the Sakhalin-2 project, and in December 2009, three major Japanese refiner companies contracted purchases of Eastern Siberia–Pacific Ocean (ESPO) pipeline crude oil cargoes from the export terminal at Kozmino Bay near Nakhodka in Russia's Far East.

At the institutional level, in 2008, the Japanese government (the Japanese Ministry of Economics, Trade and Industry's (METI) Agency for Natural Resources and Energy (ANRE)) signed a memorandum of understanding (MoU) with Russia's largest oil company Rosneft, laying the ground for energy cooperation in a number of sectors. Similar agreements were concluded later between ANRE, Itochu Corporation, Japan Petroleum Exploration Company (JAPEX), and Russia's (and one of the world's) largest energy company Gazprom in May 2009. Several important intergovernmental agreements in various sectors, including banking, nuclear power, and energy efficiency, were also signed during Prime Minister Putin's visit to Tokyo on 12–15 May 2009. Furthermore, after signing the memorandum of cooperation in the field of energy efficiency and renewable energy in May 2009, the two countries have intensified their cooperation in these areas by concluding a number of important agreements among related agencies and conducting annual bilateral seminars to facilitate information exchange and collaboration in energy efficiency and conservation.

Despite the considerable potential for Japanese-Russian energy cooperation, presently there are only a small number of projects and initiatives that have been, or are in the process of being, fully realised. Most of these bilateral projects are geographically limited to the aforementioned Sakhalin Island (the ongoing Sakhalin-1 and Sakhalin-2 projects; see Table 2 for an overview of the projects) and Eastern Siberia, including the Primorsky District (the ESPO pipeline and related projects), and focus on the development of oil and gas in these regions. Due to the remaining political hurdles, lack of mutual trust, and other outstanding problems, investment and joint projects in Russia have mainly been undertaken by

---

[3] Risa Maeda, "Japan Almost Doubles Russia Crude Imports in Aug", Fox Business, 30 September 2010.

Table 2 Overview and Current Status of Sakhalin-1 and Sakhalin-2 Projects

Name	Sakhalin-1	Sakhalin-2
Primary project developers	**US**: Exxon Neftegas (30%); **Japan**: SODECO consortium (30%); **India**: ONGC Videsh (20%); **Russia**: Sakhalinmorneftegaz (Rosneft-Sakhalinmorneftegaz Subsidiary, 11.5%), and RN Astra (Rosneft Subsidiary, 8.5%)	**Russia**: Gazprom (50% + 1), Sakhalin Energy Investment Company: Shell (27.5%); **Japan**: Mitsui (12.5%), Mitsubishi (10%)
Oil reserve estimate	2.3 billion barrels	1–1.2 billion barrels
Natural gas reserve estimate	485 billion cubic metres	507 billion cubic metres
Net total investment	US$5 billion at the initial stage; US$12 billion in total	US$4.5 billion at the initial stage; US$20 billion during the second stage over following 4–5 years
Current & expected production level	PSA signed in June 1995; the project focuses on the development of three fields — Chayvo, Odoptu, and Arkutun-Dagi. The initial stage of the project (2002–2006) focused on the development of the Chayvo field; oil production of around 50,000 barrels of oil a day started in October 2005 for delivery to the Russian Far East domestic market. Commercial oil export started in August 2006, reaching a peak rate of 250,000 barrels of oil a day in February 2007.	PSA signed in June 1994; the project has been developed in two phases. Phase 1 (1999–2003) involved oil production from an offshore platform at the Piltun-Astokhskoye field in 1999. Phase 2 (2003–2008) focused on the installation of additional platforms, construction of offshore and onshore oil and gas pipelines, an onshore processing facility, an oil export terminal and Russia's first LNG plant.

(*Continued*)

Table 2 *(Continued)*

Name	Sakhalin-1	Sakhalin-2
	Natural gas production, averaging 1.7 million cubic metres a day, began in October 2005 to meet the needs of domestic customers in the Khabarovsk Region. It reached a peak rate of 5.85 million cubic metres a day in January 2008. In December 2009, the Sakhalin-1 natural gas supplies to the Khabarovsk Region reached 5 billion cubic metres.	Commercial oil production: since December 2008 (150,000 barrels a day). LNG production: since March 2009. LNG supplies to Japan currently stand at about 5 million tonnes a year, constituting about 65% of Sakhalin-2's total LNG production. Since the LNG plant reached its full capacity (9.6 million tonnes a year) at the end of 2010, it is currently producing about 5% of global LNG supplies.
Target markets & consumers	Following the signing of the MoU on gas supplies with China's CNPC in October 2006, Exxon was at odds with the rest of the project developers over the Sakhalin-1 gas export strategy. While Exxon preferred pipeline exports to China (as well as Japan and Korea), other shareholders (as well as Gazprom) insisted on giving the priority to piping to the LNG terminal at Sakhalin-2 or to the domestic market in the Russian Far East (particularly the Khabarovsk Region). The issue of gas marketing and export routes from Sakhalin-1 was partially resolved in May 2009 when the project shareholders came to an agreement to sell 20% of its gas production starting in 2012 to Gazprom for delivery to the Russian domestic market.	Oil and LNG exports: primarily Japan, other markets in Asia (Korea), as well as North American (including Hawaii and Mexican west coast) markets.

(Continued)

Table 2 (*Continued*)

Name	Sakhalin-1	Sakhalin-2
Status/notes	In May 2009, the next phase of the Sakhalin-1 project was signified with the start of drilling at the Odoptu field; commercial production is expected to begin at the end of 2010. In July 2010, oil production stood at over 270 million barrels, while natural gas deliveries to the domestic market in the Russian Far East reached 6 million cubic metres. The future phases of the project focus on the development of Chayvo gas reserves for export as well as development of the Arkutun-Dagi oil and gas field (expected to begin production in 2014).	On 29 March 2009, the first LNG cargo from Sakhalin was loaded from the Sakhalin-2 LNG plant for delivery to Japan's Tokyo Gas and Tokyo Electric companies. In October 2009, as additional US$1.4 billion was secured in project financing in order to advance the drilling programme of Phase 2 of the project.

Source: The Sakhalin-1 project website (http://www.sakhalin1.com/Sakhalin/Russia-English/Upstream/about.aspx) and Sakhalin Energy website (http://www.sakhalinenergy.com/en/ataglance.asp?p=aag_main&s=14).

relatively large Japanese firms and corporations capable of securing Japanese government financial and diplomatic support.

3.3.1 Sakhalin Projects

Historically, Sakhalin oil and gas development has not only been an attractive area for Japanese business involvement, but also the Japanese government (the projects are among the most extensive and largest commitments made by the government). For the Sakhalin-1 project, Japan is represented by a consortium made up of the Sakhalin Oil and Gas Development Company (SODECO), along with Itochu and Marubeni corporations, holding a 30% share. The Japanese members of the Sakhalin-2 consortium are the Mitsui Bussan and Mitsubishi Shoji corporations, holding 12.5% and 10% of the project shares, respectively. Both Sakhalin-1 and Sakhalin-2, which are among the world's largest oil and gas integrated projects, hold extensive recoverable gas and oil reserves, the development of which requires tens of billions of US dollars in investment based on production-sharing agreements between the Russian government and participating businesses.

The Japanese government has strongly supported Japanese business participation in both projects. It has also been actively involved in providing financial assistance to the Sakhalin-2 project, focusing on the construction of the first Russian LNG terminal located in southern Sakhalin, signalling the growing significance of Russian LNG in Japan's energy import diversification strategy. According to Sakhalin Energy's statement on 16 June 2008, the Japanese Bank for International Cooperation (JBIC) along with a banking consortium of international (mainly Japanese) banks agreed to provide a US$5.3 billion financial package. The funds helped finance the final stages of the LNG construction project in Prigorodnoye, located in Aniva Bay of the Korsakov District of the Sakhalin Region.

Russian President Dmitry Medvedev and then Prime Minister of Japan Taro Aso both attended the ceremony launching Russia's first LNG plant on 18 February 2009, thereby underlying the significance of the project to the two countries and the Asia-Pacific region as a whole. At the opening ceremony, the two leaders pledged to apply a new, "innovative and unconventional" approach to solve the territorial dispute in a timely manner, while continuing to expand their ties in the energy, infrastructure, and environmental sectors. However, it is important to note that the Sakhalin Region holds an administrative jurisdiction over the disputed Kuril Islands and is known to be a region with a highly unyielding public opinion on the territorial dispute. Aso's visit, which was the first visit to Sakhalin by a Japanese leader since WWII, stirred a lot of controversy among Japanese conservatives pushing for the revival of the *seikei fukabun* policy toward Russia.

Despite the global financial crisis, falling energy prices, and falling demand that have cast a dark shadow on Moscow's energy ambitions in the region, there are still fears in Tokyo that Russia may gain the upper hand vis-à-vis Japan, by expanding its role in the gas and LNG markets in Japan and the Asia-Pacific region as a whole. Some raise concerns that Russia, while asserting its role in the region and expanding its energy relations with Japan, has been effectively focusing only on the economic aspects of bilateral cooperation, thereby pushing the Kuril territorial dispute off the negotiation table and hence hindering the normalisation of political relations with Japan. In response to the Japanese voters' conservative sentiment, Aso reiterated his commitment to pursue the solution to the territorial dispute based on Japan's official policy of insisting on the attribution of all four islands. During a press conference on the expected results of the May 2009 Japanese-Russian summit meeting, then Japanese Foreign Minister Hirofumi Nakasone stressed, "from Japan's perspective, the most important issue between Japan and Russia [was] the issue of the Northern Territories… and the conclusion of the peace treaty"; thus it was critical for the Japanese government to advance its negotiations on the territorial dispute to keep the level of importance in line with the negotiations on bilateral economic cooperation.[4]

Despite the remaining political problems and concerns, in early 2009, Russia successfully realised its entry into the world and Asian LNG market. LNG production from the Sakhalin-2 project at the new terminal began on 5 March, with the first shipment of LNG (4.8 million tonnes)

[4] Japanese Ministry of Foreign Affairs (MOFA), Press Conference by Minister for Foreign Affairs Hirofumi Nakasone, 12 May 2009, www.mofa.go.jp/announce/fm_press/2009/5/0512.html [Accessed 14 October 2009].

reaching Japan in early April 2009. According to Sakhalin Energy's official statement, as of late December 2010, the LNG plant has reached its full production capacity of 9.6 million tonnes a year, constituting 5% of the world's LNG market and about 4.3% of Japan's LNG imports. About 65% of the total capacity has been contracted on a long-term basis to eight Japanese companies, including Tokyo Gas (TOGAS) and Toho Gas, and the remaining 35% to the South Korean and US markets. It has been recently reported that the Japanese government is planning to increase purchases of Russian LNG in 2011 to help diversify its LNG imports.[5]

New opportunities for bilateral cooperation in this region include the Gazprom-led Sakhalin-3 project. Given the success of their partnership with Gazprom in the Sakhalin-2 project, Gazprom CEO Alexei Miller, during his visit to Japan in 2009, personally invited Japan's Mitsubishi and Mitsui corporations to take part in the Sakhalin-3 and possibly Sakhalin-4 projects on a "financial, technological, and to some extent, operating level", stressing that Gazprom's existing partners would be given an advantage.[6]

Another possible energy-related joint project involves the construction of the 1,350 km Sakhalin-Khabarovsk-Vladivostok gas pipeline, which commenced in July 2009 and is expected to operate at capacity by 2011. According to recent reports, Japanese steel producers, through the JBIC, have been negotiating a financial assistance package to Russia (reportedly a US$1 billion commodity loan tied to the supply of about 300,000 tonnes of Japan-produced steel pipes) for the construction of the pipeline.[7] Upon its completion, the pipeline will transport Sakhalin's offshore natural gas to the Vladivostok terminal for deliveries to Japan, China, and other consumers in the Asia-Pacific region. In addition, in July 2010 the Japanese and Russian governments reached a preliminary agreement on the construction of a new LNG terminal near Vladivostok in Russia's Far East in order to deliver LNG from Sakhalin and Eastern Siberia to Japan and

[5] "Japan Mulls Hiking LNG Imports from Russia in 2011", IHS Global Insight, 5 October 2010.

[6] "Russia: Gazprom Mulling Invitation to Foreign Partners for New Sakhalin Projects", IHS Global Insight, 3 September 2009.

[7] See also *Russia and CIS Oil and Gas Weekly*, no. 34 (27 August–2 September 2009), p. 15; and IHS Global Insight, 3 September 2009.

other Asian customers. JAPEX, jointly with Gazprom, have also launched a feasibility study on the construction of a related gas chemical plant and LNG facility in the region. By assisting Russia in these infrastructure projects, the Japanese government would help to boost gas supplies to the Pacific Coast in the Russian Far East, thereby ensuring shipments to Japan and other markets in the Asia-Pacific region.

Japanese companies, such as Mitsui Bussan Corporation, have shown interest in joining Gazprom in the development of gas fields not only in Eastern Siberia, for example the Chayandinskoye gas field with estimated gas reserves of 1.2 trillion cubic metres (Tcm) of gas, but also gas fields including LNG projects in the European part of Russia. One such project is the Gazprom-led Shtokman project, which includes the development of a giant gas field in the Barents Sea and the planned construction of a 7.5-million-tonne-per-year LNG plant. Another project is the Nord Stream gas pipeline project, which would link Russia and Germany and would help transport Russia's natural gas to the rest of Europe. In late January 2010, Japanese Sumitomo won a tender, along with Russia's United Metallurgical Company (OMK) and Germany's Europipe, to supply pipes (10% of 1 million tonnes in total) for the second stage of this project.[8]

3.3.2 The Eastern Siberia–Pacific Ocean Pipeline and Related Projects

In addition to the Sakhalin projects discussed earlier, there are a number of prospective joint projects that focus on the development of energy resources in Eastern Siberia and Russia's Far East. One of these projects is the construction of the Eastern Siberia–Pacific Ocean pipeline and development of the related regional infrastructure required to bring Russian oil to the Asia-Pacific market (Figure 2).

This large-scale project was divided into two construction phases. Phase 1, which focused on the construction of the 2,757 km Taishet-Skovorodino branch to deliver around 600,000 barrels of oil a day, was completed and became operational on 28 December 2009. Prime Minister Putin,

[8] "OMK, Europipe Sumitomo to supply pipes for phase two of Nord Stream," *Russia and CIS Oil and Gas Weekly*, no. 3 (920) (21 January–27 January 2010), p. 26.

Figure 2 Eastern Siberia–Pacific Ocean Pipeline

Source: "Russia turning to China's huge energy thirst." The Telegraph Online Supplement by Rokiyskaya Gazeta, October 29, 2010. <http://www.telegraph.co.uk/sponsored/russianow/business/8097658/Russia-turning-to-Chinas-huge-energy-thirst.html>. Accessed on December 14, 2010.

who attended the ESPO pipeline and oil terminal launching ceremony, stressed the critical strategic importance of this multi-billion-dollar project. The shipment of the first crude supply from the new oil terminal at Kozmino Bay (the final point of the ESPO pipeline near Nakhodka City, Primorsky District) took place in January 2010, thereby officially launching ESPO Blend exports to the Asia-Pacific market.

In February 2010, a joint venture between trading house Mitsui and ExxonMobil won a bid for the delivery of a 730,000-barrel ESPO crude cargo to Japan's refineries.[9] Upon the completion of Phase 2 of the ESPO pipeline project in 2014, which would connect Skovorodino with Russia's Pacific Coast (about 2,100 km), crude oil from the Siberian fields will be delivered to the Kozmino Bay terminal at a total capacity of 1.6 million barrels a day. This project, which mainly targets Russia's East Asian customers, is also considered indispensable for the economic development of Russia's Far East.

[9] "Russia to Export East Siberian Crude Oil to Japan", IHS Global Insight, 26 February 2010.

Japanese energy companies have long been eyeing a direct participation in the development of Russia's natural resources in Siberia and the Far East. In late April 2008, Japan's state-run Oil, Gas and Metals National Corporation (JOGMEC) obtained access to Russian oil resources in Eastern Siberia by acquiring the rights to develop the Severo-Mogdinsky oil and gas block in the Irkutsk Region (the block, with a total area of 3,747 km^2, is located 150 km southeast of the planned route for Phase 2 of the ESPO pipeline). The Japanese side agreed to explore and develop the acquired block together with Russia's Irkutsk Oil Company (IOC) by establishing a joint venture — IOC-Sever — and jointly investing US$95.8 million in this project.[10] Under the agreement, JOGMEC will provide the latest technology, in particular for seismic studies, exploration, and development of the block. In June 2009, it was reported that the drilling and assessment work in the block had successfully begun.

In addition, on 1 September 2008, JOGMEC and Russia's United Oil Group Ltd. (UOG) signed a joint venture agreement (in which UOG will have 51% and JOGMEC 49%) to implement projects in the sphere of oil and gas prospecting, exploration, and production in Russia. The priority areas for this project are Russia's Krasnoyarsk Territory, Irkutsk Region, and Sakha Republic (Yakutiya). The parties announced in May 2009 that a joint five-year feasibility study and prospecting work had begun. It expected that the crude oil production in this area would be tailored mostly to satisfy the demand of the Japanese market.

JOGMEC's presence in the Russian Far East energy market further expanded after signing another important agreement during Prime Minister Putin's visit to Japan in May 2009. According to this agreement, JOGMEC and IOC established a new joint venture IOC-Zapad and agreed to explore the two additional blocks of Bolshetirsky and Zapadno-Yaraktinsky in the same region. By taking part in these upstream projects,[11] Japanese energy-related businesses not only aspire to boost their

[10] Following the Putin-Abe Summit in June 2007, the Irkutsk Oil Company and JOGMEC formed a 51:49 joint venture (INK-Server Ltd.) in order to launch joint development of the Severo-Mogdinsky oil and gas block.

[11] The upstream sector refers to the exploration and production (E&P) sector of energy operations. It involves searching for, recovering, and producing crude oil and natural gas.

operations in the Russian energy market, but also to ensure the Russian government's commitment to the timely and successful completion of Phase 2 of the ESPO pipeline.

Japan and Russia have recently discussed a number of projects for Eastern Siberian downstream sector development.[12] One project (discussed during the Governor of Primorsky District Sergei Darkin's visit to Japan on 12 May 2009) involved the construction of Rosneft's planned 200,000- to 400,000-barrel-per-day refinery near the Kozmino Bay terminal. Located at the final destination point of the ESPO pipeline, this large-scale refinery would provide many economic benefits to the Russian Far East.

Gazprom has also been considering a number of projects to build gas chemical facilities in Eastern Siberia to complement the development of the area's gas resources. These large-scale projects would most likely require multilateral participation of foreign partners, including Japanese companies that have already shown interest in a joint project focusing on the construction of a gas chemical plant in the Republic of Sakha (Yakutiya) and other areas in Eastern Siberia.[13] In November 2009, the IOC and JOGMEC signed an agreement to prepare the feasibility study for the creation of gas-to-liquid (GTL) capacity in Russia, particularly in Eastern Siberia and the Russian Far East.[14] Such projects will utilise the investment and technological potential of Japanese partners, while providing them with access to the Russian downstream sector and other possible benefits of joint development of energy resources in this region.

Apart from the oil and gas sectors, Japan and Russia have also recently become engaged in bilateral negotiations on enhancing their cooperation in other fields, such as nuclear and renewable energy resources as well as environmental protection, energy efficiency and conservation. Since Russia lacks experience in these areas, except the nuclear sector, Japan could contribute its expertise, advanced technology, and know-how.

---

[12] The downstream sector refers to the refining of crude oil as well as the sale and distribution of natural gas and products derived from crude oil (such as liquefied petroleum gas (LPG), gasoline or petrol, jet fuel, diesel oil, other fuel oils, asphalt and petroleum coke).

[13] "Japan, Russia's Rosneft Agree on Energy Cooperation", Reuters UK, 21 March 2008.

[14] *Russia and CIS Oil and Gas Weekly*, no. 47, (26 November–2 December 2009), p. 15.

In the nuclear energy sector, Japan's Toshiba Corporation entered into a joint agreement with Russia's state-owned Atomenergoprom on 20 March 2008 to promote bilateral civilian nuclear energy projects, including joint uranium enrichment initiatives. According to the Director General of Atomenergoprom, Sergei Kiriyenko, Japanese-Russian energy cooperation in the nuclear sector would be "beneficial not only to the employees of [the two] companies, but also to users of products and services related to [the] nuclear cycle throughout the world".[15] The Toshiba press release announced that the agreement would contribute to the "stable and secure supply" of nuclear fuel cycle services in Japan, the US, and elsewhere. Furthermore, it would also help strengthen the "complementary relations" that would "lead to the establishment of a bilateral strategic partnership".[16]

Starting in early in 2009, the Russian state nuclear corporation, Rosatom (which is the de facto Russian agency for nuclear energy and has been looking to increase Russia's participation in the growing nuclear energy market in the Asia-Pacific region) was in negotiations with the Japanese government about signing a joint cooperation agreement in the field of nuclear energy. A joint agreement, which was finally concluded during Russian Prime Minister Putin's visit to Japan in May 2009, provided for the delivery of Russian supplies of low-enriched uranium to Japan and the construction of a depot for its storage on Japan's territory, while Japan would supply Russia with advanced nuclear power plant technology in exchange. It also envisioned increasing the share of Russia's presence in Japan's nuclear energy market from its current 15% to 25% in the near future.

In addition to bilateral cooperation in the nuclear energy field, during Prime Minister Putin's visit to Tokyo in May 2009, METI and the Russian Ministry of Energy signed an MoU to boost their cooperation in the field of energy efficiency and renewable energy at the institutional

---

[15] "Toshiba, AtomEnergoProm Sign Framework Agreement", World Nuclear News, 20 March 2008.

[16] Toshiba Corporation, Press Release, "Toshiba Signs General Framework Agreement with Atomenergoprom to Explore Collaboration in Civilian Nuclear Power," 20 March 2008, http://www.toshiba.co.jp/about/press/2008_03/ pr2001.htm, accessed on 12 December 2010.

(government) and private-sector levels, including the establishment of the International Partnership for Energy Efficiency Cooperation (IPEEC) and participation in a number of joint research projects and capacity-building initiatives. The MoU stressed that the two countries' cooperation in these sectors could help stabilise the overall energy balance not only in Japan and Russia, but also in the Asia-Pacific region as a whole.[17] In addition, the newly established Russian Energy Agency (REA) has been actively involved in a number of joint activities for the promotion of energy efficiency and energy conservation policies with Japanese public and private organisations, including Japanese Business Alliance for Smart Energy Worldwide (JASE World); JBIC; JETRO; the Institute of Energy Economics, Japan (IEEJ); Energy Conservation Center, Japan (ECCJ); and others. On 27 October 2010, JETRO — in partnership with JBIC, REA, and Russia's Sberbank — organised a joint seminar in Moscow on "Possibilities for Enhancing Japanese-Russian Environmental Cooperation".

3.4 PROSPECTS FOR FUTURE JAPAN-RUSSIA BILATERAL ENERGY COOPERATION

Despite the remaining obstacles and political problems, Russia and Japan are moving toward a new stage in their relations. The bilateral economic ties and trade relations have been dramatically improving, and the two countries have been expanding their cooperation in various sectors of the energy field. The governments of the two countries have come to realise that further development of complementary economic relations with each other, focusing on energy development in Russia's Eastern Siberia and Sakhalin, as well as improvement of mutual ties in all other areas, is critically important to both Russia and Japan. In addition to the realisation of the mutually beneficial aspects of improved bilateral relations, another factor that could bring the two countries closer together is the political and economic expansion of China in the Asia-Pacific region, which has caused

[17] The full text of the MoU is available at the Ministry of Energy of Russian Federation webpage (in Russian), http://minenergo.gov.ru/documents/soglasheniya/1145.html [Accessed 20 October 2009].

growing concerns in both Japan and Russia. As such, full normalisation of Japanese-Russian relations is a welcome development that would greatly contribute to the economic and political stability as well as energy cooperation and security of the entire Asia-Pacific region.

The Russian Energy Strategy to 2030 adopted by the Russian government in November 2009 stressed the importance of energy cooperation with many foreign economic partners, including Japan, which represent a lucrative export market for Russian crude oil, nuclear fuel, natural gas, and LNG, as well as petrochemical products. Presently Russia views its political presence in the Asia-Pacific region (via APEC; the annual meeting in 2012 will be held in Vladivostok) as very important and regards the Asia-Pacific market as one of the most attractive due to its growing energy demand (including crude oil and LNG), stable energy prices, great potential for growth, and other business and cooperation opportunities in the field of energy.

The most recent successful developments in bilateral energy cooperation were the inauguration of the Prigorodnoye gas liquefaction plant in late March 2009 under the Sakhalin-2 project and the December 2009 launch of the Kozmino Bay oil terminal under the ESPO pipeline project. They were important milestones that helped bring Japanese-Russian energy cooperation to a new level, marking the beginning of large-scale energy exports from Russia to Japan and opening Russia's access to Asia-Pacific markets.

The growing interest within Japanese business circles in ongoing and planned projects in Russia's Far East and Eastern Siberia underscores the view that the energy resources of these nearby regions could serve as an important strategic reserve for Japan in its quest to secure long-term supply sources. In addition, a number of new opportunities for potential bilateral cooperation have recently presented themselves with the two countries adopting more assertive policies in the fields of energy conservation and efficiency, promotion of nuclear and renewable energy resources, and advancement of clean energy technologies with the aim to address worldwide concerns about climate change and environmental protection.

The growing Japan-Russia economic ties and energy cooperation have been welcome developments, given that the two countries' relations have been historically tainted by their outstanding political problems. Today, the international circumstances as well as present political and economic situation in the two countries' relations seem to favour the development and feasibility of their energy cooperation. Unlike during the ideology-driven Cold War, the problems related to the acquisition of funds for project financing or technological assistance are no longer subject to political considerations or the objections of strategic allies. However, the two countries are presently facing the necessity of building political trust with each other and focusing on finding mutually acceptable solutions to their remaining hurdles.

Although ideological disagreements no longer seem to affect Japanese-Russian cooperation, political tensions (including the unresolved territorial dispute) and a lack of clear and constructive long-term policies toward each other continue to cast a shadow over their economic relations, including their energy cooperation. As Dmitri Trenin and Vasily Mikheev of the Moscow Carnegie Center point out in their report, "Russian and Japanese leaders need to recognise that not having any form of a solution to the territorial problem could harm both countries, at the very least in the form of failed opportunities and untapped potential".[18]

Thus, despite expanding their bilateral trade and energy cooperation, Japan and Russia are yet to maximise and exploit the fruits of their full-fledged energy cooperation. The remaining problems and concerns may delay or put the feasibility of the proposed joint development initiatives and projects at risk. In order to boost the chances for the realisation of bilateral energy projects, the following issues, in addition to the outstanding political problems such as the territorial dispute, have to be addressed as effectively as possible:

- **Short- to medium-term issues**: Russia's prohibitive tax regime, lack of transparent business environment and effective regulatory regime; shortage of large-scale capital investment and other project financing problems; shortage of necessary technology and equipment as well as a skilled labour force; lack of infrastructure, difficult weather/climate conditions in Russia, and great distances.

[18] D. Trenin and V. Mikheev, "Russia and Japan as a Source for Mutual Development: A 21st -Century Perspective on a 20th-Century Problem" (Carnegie Endowment for International Peace, Moscow Center, 2005), p. 19.

- **Long-term issues**: Russia's endemic corruption, prohibitive tax regime, and lack of efficient structure in energy strategy (especially vis-à-vis foreign investors); domestic and international political tensions (possibly for both Russia and Japan); Japan's cautious business tactics in Russia; the rising cost estimates of joint projects; energy market fluctuations (such as staggering demand and/or inefficient supply); and others.

Notwithstanding the fact that Russia would greatly welcome Japanese investment and technology for the development of Sakhalin and Siberian energy resources, the Japanese government, citing a lack of progress on the Kuril territorial dispute, has been displaying a cautious attitude toward increased participation by Japanese firms in Sakhalin and expansion of their partnership with Gazprom, thereby adhering to a wait-and-see approach for the time being.

On 25 November 2009, it was reported that the Institute of Marine Geology and Geophysics of the Russian Academy of Sciences (Far Eastern branch) had recently published a report on the geological potential of the oil and gas reserves in the Middle Kuril curve, which stretches between the disputed islands of Shikotan and Kunashir. While the news of potentially large oil and gas reserves (estimates stand at 1.2–1.6 billion tonnes of fuel equivalent[19]) are welcome in Russia, particularly in the Sakhalin region, it may raise many concerns in Japan that the likelihood of the return of the disputed islands may now be greatly undermined. While the report strongly recommended the Russian government look seriously at the matter, if the two countries could move beyond their political contentions, this discovery might present a unique opportunity for them to put their efforts together for a joint geological study and future development of the Kuril resources.

Furthermore, President Dmitry Medvedev's brief visit to Kunashir (one of the disputed Kuril Islands) on 1 November 2010 only added to the existing bilateral tensions. This visit had a special significance not only because it was the first official visit to the disputed territory by the top leader in the history of the Soviet Union and Russia, but also because,

---

[19] See *Russia and CIS Oil and Gas Weekly*, no. 46, (19–25 November 2009), p. 11.

while in Kunashir, President Medvedev announced that the Russian government would provide necessary investment to boost local economic development in the Kuril Islands. While the Russian government insisted that there was no connection between the President's visit to Kunashir and Russian-Japanese relations, the Japanese government reacted immediately by issuing an official protest and even temporarily recalling the Japanese ambassador from Moscow. The conservative circles in Japan, citing the Japanese government's inability to reach a common understanding with Russia on the territorial issue, have begun pressuring the government to review Japan's economic relations with Russia until there is a breakthrough on the territorial issue.[20]

Notwithstanding the ongoing tensions in their political relations, the top leaders in both Japan and Russia managed to prevent the recent conflict from escalating. While "agreeing to disagree" on their respective positions in the Kuril dispute, they chose to focus on future mutual benefits for both nations. When Prime Minister of Japan Naoto Kan met with the Russian President on the sidelines of the recent APEC summit meeting in Yokohama on 13 November 2010 (Figure 3), the two leaders reiterated their will to build a stable trust-based relationship between their countries and resolved to boost their economic cooperation, which in turn would help improve their political relations focusing on the Kuril territorial dispute. They also agreed to hold a summit meeting early next year in Russia to continue their bilateral talks for the conclusion of a peace treaty.[21]

The search for solutions to the aforementioned outstanding problems requires effective and timely issue-specific approaches introduced at the government level in Japan and Russia. Furthermore, in order to effectively overcome the obstacles and minimise the associated risks, it is essential for the two countries' leaders, through their political will and determination, to continue their diplomatic dialogue, while striving to expand bilateral relations in all spheres. If they succeed in their efforts, there would be a greater chance for full-fledged Japanese-Russian cooperation that would greatly benefit both countries. In this respect, Japanese-Russian energy

---

[20] Yuka Hayashi and Gregory L. White,"Visit Pits Japan, Russia", *The Wall Street Journal*, 2 November 2010.

[21] "Kan, Medvedev Discuss Disputed Isles", *Japan Today*, 13 November 2010.

Figure 3 Japanese PM Kan and Russian President Medvedev at the November 2010 APEC Meeting in Yokohama, Japan.

Source: "Meeting with PM Naoto Kan: APEC Summit in Yokohama." Official Website of the Russian President. November 13, 2010. <http://www.kremlin.ru/visists/9506>. Accessed on December 14, 2010.

collaboration has a lot of potential based on complementary needs realised in a number of ongoing and prospective projects. Russia and Japan are important players in the world energy market, particularly in the Asia-Pacific region. A successful energy collaboration between the two countries would not only serve their respective economic and energy needs, but would also improve their bilateral ties as a whole. Finally, it would contribute greatly to the strengthening of energy security and cooperation in the Asia-Pacific region.

BIBLIOGRAPHY

Bierman, Stephen, and Pronina, Lyubov. "Russia, Japan Should Boost Trade Ties, Medvedev Says", Bloomberg, 17 February 2009.

JETRO. *Japan's 2009 Trade Statistics*, http://www.jetro.go.jp/world/japan/stats/trade/ [Accessed 12 December 2010].

Ministry of Energy of the Russian Federation. *Russian Energy Strategy for the Period up to 2030* (Moscow: Institute of Energy Strategy), http://energystrategy.ru/projects/docs/ES-2030_(Eng).pdf [Accessed 16 December 2010].

"Vladivostok Pipeline", *Russia and CIS Oil and Gas Weekly*, no. 34 (900), (27 August–2 September 2009).

"Irkutsk Oil CO, JOGMEC to prepare GTL feasibility study in 2010", *Russia and CIS Oil and Gas Weekly*, no. 47 (913), (26 November–2 December 2009).

Sakhalin Energy. http://www.sakhalinenergy.com/en/ataglance.asp?p=aag_main&s=14 [Accessed 12 December 2010].

"South Kurils Shelf Rich in Oil, Gas — Scientists", *Russia and CIS Oil and Gas Weekly*, no. 46 (912), (19–25 November 2009).

The Sakhalin-1 Project. http://www.sakhalin1.com/Sakhalin/Russia-English/Upstream/about.aspx [Accessed 12 December 2010].

CHAPTER 4

ENERGY-RELATED POLICY ISSUES IN TERMS OF JAPAN-CHINA RELATIONS

Yuji Morita

4.1 INTRODUCTION

The People's Republic of China celebrated its 60th anniversary on 1 October 2009. It was in October 1949, 60 years ago, when Mao Zedong announced the founding of China at the Tiananmen Square in Beijing. The first 30 years after the founding of China were marked by many political conflicts caused by the Cultural Revolution initiated by Mao Zedong with the aim of establishing a socialist system and thereby increasing productivity. As a result, the country became ravaged instead. With the death of Mao Zedong in September 1976 and the end of the Cultural Revolution, a new political direction was adopted with an emphasis on economic growth.

The reform and opening-up policies initiated by Deng Xiaoping, a realist who consolidated his leadership in December 1978, virtually abandoned socialist ideas. In order to rebuild the national economy ruined by the Cultural Revolution, China began exporting its vast energy resources as a means of obtaining foreign currencies. Against this background, oil accounted for approximately 20% of its total exports in the early 1980s. At the same time, China promoted opening-up and took measures to build its domestic industry.

For Japan, which needs to import vast quantities of energy from overseas, the first oil crisis was an event that increased the importance of China as a new oil supply source. It took place immediately after Japan had established diplomatic relations with China in 1972 following the end of the Second World War in 1945 and the Korean War in 1953. In 1979, Japanese oil companies started oil exploration and development activities in China to look for Chinese oil resources.

In China, as a result of the belt-tightening policies adopted to restrain rapid liberalisation, the Tiananmen Incident took place in June 1989 in which the army killed students and civilians. However, thereafter, reform and opening-up continued. The dissolution of the Soviet Union in 1992 provided an opportunity for China to pursue a market economy. With an expansion of scope of the opening-up policies, together with a surge in direct investment from overseas, China entered into a new era and came to enjoy double-digit economic growth. The gross domestic product per capita rose from $190 in 1978 to $3,266 in 2008.

With economic development, energy consumption also increased. Although its domestic oil resources were being explored and developed with utmost effort, the demand for oil came to exceed its domestic production. Its scarce natural gas resources could not make up for the lack of domestic oil supply. As a result, China became a net oil importer in 1993. As for coal, domestic coal resources are abundant, but the transport capacity of the railway to carry coal to the market was poor. Therefore, it became necessary to invest and expand the railway transport infrastructure.

Under these circumstances, it became a critical issue for China to ensure energy resources — stable oil supply in particular — to sustain its economic development. At about the same time, the Japanese Prime Minister paid a visit to the Yasukuni Shrine, where those soldiers who died in the Second World War are commemorated and worshiped. This resulted in the rapid deterioration in Japan-China relations. Japan and China competed against one another to secure energy resources abroad and had turbulent relations with occasional serious confrontations.

In the August 2009 general election, the Democratic Party of Japan (DPJ) defeated the long-governing Liberal Democratic Party (LDP). With the change of the Japanese government, relations between Japan and China improved once again. It seems that both countries are trying to promote energy savings through technical cooperation and looking for ways to consolidate the bilateral relations in the international fight against global warming.

4.2 CHINA'S ENERGY DEMAND

4.2.1 Energy Consumption in China

Energy consumption in China is increasing sharply, reflecting its high economic growth in recent years. Its average annual growth rate from 1990 to 2007 was 5.9%, which substantially exceeded the world average of 1.9%. The annual growth rate from 2000 to 2007 was 10.2% and the amount of its energy consumption reached 1,761 million tonnes of oil equivalent (Mtoe) in 2007. China became the second-largest energy consumer in the world, after the United States (Figure 1).[1,2]

The breakdown of energy consumption by source indicates that coal had a very large share of total energy consumption in 2007, i.e. 73.0%. Among other energy sources, oil accounted for 20.2%, natural gas for 3.4%, nuclear energy for 0.9%, and hydropower for 2.4%. The amount of natural gas consumption was not substantial. In terms of the growth rates since 1990, China has seen a high growth in the consumption of oil, natural gas, and hydropower — 7.1%, 9.4.%, and 8.2%, respectively — against the 5.4% growth in coal consumption. Since its introduction in 1993, the

---

[1] On the other hand, energy consumption in Japan has remained more or less unchanged for several years, due to the economic slowdown and as a result of the implementation of energy-saving measures. The average annual growth rate from 1990 to 2007 was 0.9%. The annual growth rate since 2000 has been –0.1%.

[2] 1 Mtoe equals 1 million tonnes of oil. 1 toe = 1×10^7 kcal. The data used in this paper are primarily from the *Energy Balances of Non-OECD Countries and Energy Balances of OECD Countries* compiled by the IEA. The data, however, do not include non-commercial energy consumption in non-OECD countries.

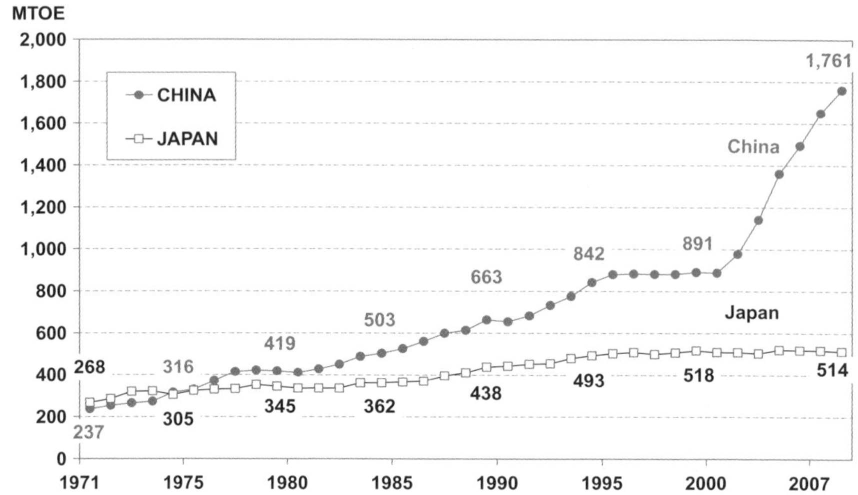

Figure 1 Primary Energy Consumption in China and Japan

Source: IEA, *Energy Balances of Non-OECD Countries; Energy Balances of OECD Countries.*

consumption of nuclear power has increased at a rate of nearly 30% per year (Figure 2).[3]

4.2.2 China's Energy Output

With regard to China's energy production, it produced 32 million tonnes of coal in 1949. The oil output was merely some 120,000 tonnes (2,400 barrels per day). In 1978, when the reform and opening-up began, China produced 618 million tonnes of coal and 104.05 million tonnes of oil (2.09 million barrels per day). Since then, its energy production had increased to produce 2.793 billion tonnes of coal and 190.02 million tonnes of oil (3.82 million barrels per day) in 2008. Production of coal, oil, and natural gas in 2008 was 4.5 times, 1.8 times, and 5.9 times the production in 1978, respectively. Capacity of the electric power facilities and the production of electricity had also increased by 13.9 times and 13.5 times,

[3] The breakdown of energy consumption by source in Japan indicates that oil accounts for 44.8%, coal for 22.3%, natural gas for 16.2%, nuclear power for 13.4%, and hydropower for 1.2%. The growth rate since 1990 in consumption of oil was –0.5%, coal 2.5%, natural gas 3.8%, nuclear power 1.6%, and hydropower –1.1%.

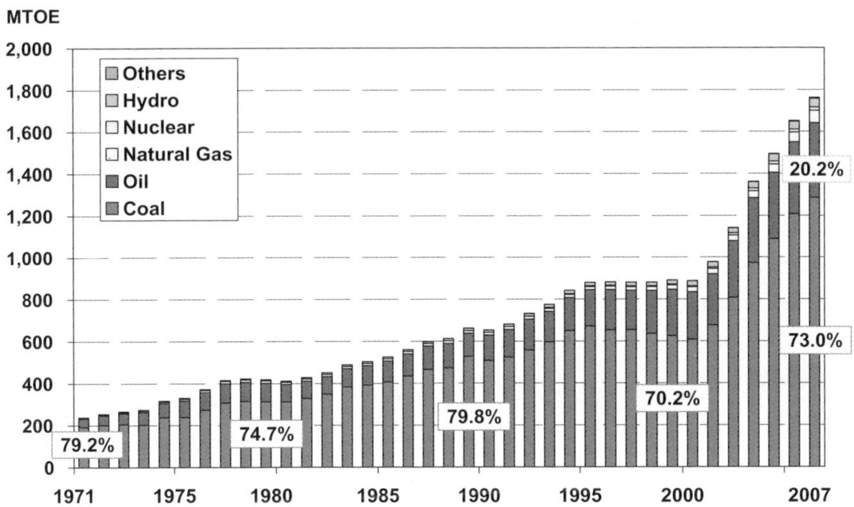

Figure 2 China's Primary Energy Consumption by Energy Type
Source: IEA, *Energy Balances of Non-OECD Countries; Energy Balances of OECD Countries.*

Table 1 China's Energy Production

		1949	1978	2008
Coal	Million Ton	32	618	2,793
Oil	Thousand Ton	120	104,050	190,020
	Thousand B/D	2.4	2,090	3,820
Natural Gas	Million m3	10	13,730	80,510
Power Generation Capacity	GW	1.85	57.12	792.53
Electricity Production	Twh	4.3	256.6	3,466.9

Source: National Bureau of Statistics of China, *China Statistical Yearbook*; Statistical Communiqué of the People's Republic of China on the 2008 National Economic and Social Development, 26 February 2009; and others.

respectively. This indicates that an increase in energy output corresponds to the growth of the economy (Table 1).[4]

[4]On a global scale, China is the fifth-largest oil producer after Saudi Arabia, Russia, the United States, and Iran, and it is the biggest coal producer. China is also on the list of the 10 biggest natural gas producers.

4.2.3 Energy Supply and Demand

If a country's self-sufficiency ratio is to be calculated by dividing domestic energy output by domestic energy consumption, China's energy self-sufficiency rate became less than 100% in 2002 and continued to decline since then to 92% in 2007. In other words, it has had little energy export capacity in recent years. Instead, China has become dependent on import for certain energy sources. Although it produces sufficient coal to meet domestic demand, China depends on imports for oil and natural gas. As at 2008, China is 53% self-sufficient in oil and 98% self-sufficient in natural gas (Figure 3).

4.3 CHINA'S DEMAND FOR OIL

4.3.1 Development of Domestic Oil Resources

The biggest slogan in the exploration and development of domestic oil resources in China is "Stabilise the East, Develop the West". Most of the oil output is from oil fields in the eastern part of China, e.g. Daqing,

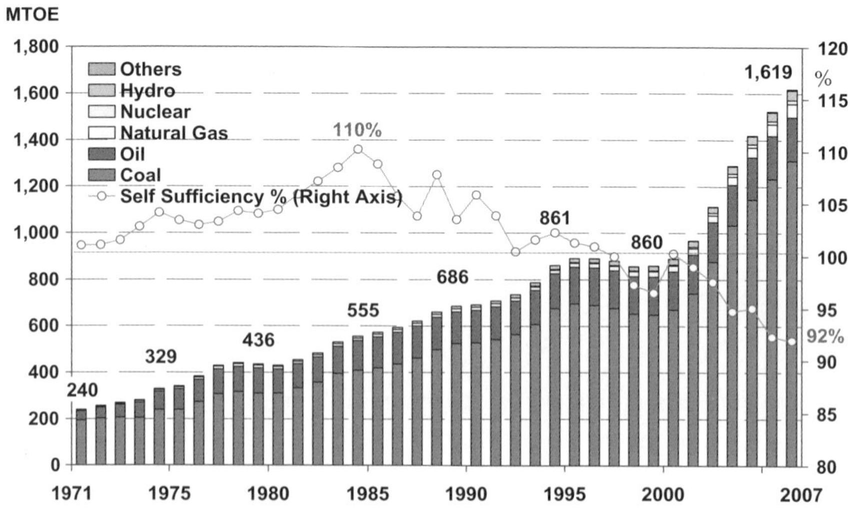

Figure 3 China's Energy Production

Source: IEA, *Energy Balances of Non-OECD Countries.*

Liaohe, and Shengli Oil Fields. However, as these major fields have been in operation for quite a number of years, their output capacity has already passed their peak. Thus, China is struggling to maintain the productivity and to extend the life of these mature oil fields.[5]

Exploration of oil resources in the western part of China started in the 1950s. Though exploration activity began with assistance from the former Soviet Union, it was fatally dashed in the face of the deterioration of Sino-Soviet relations and the Cultural Revolution in the 1960s. During the period from 1983 to 1988, full-scale exploration work was resumed, with a focus on the development of resources in the three basins in Xinjiang Uyghur Autonomous Region, i.e. Jungar, Tarim, and Turpan Basins, as well as in the Ordos Basin in Gansu Province. In 2008, the output of China National Petroleum Corporation (CNPC) at Xinjiang Oil Field (in the Jungar Basin), Tarim, Turpan-Hami, and Changqing Oil Fields (in the Ordos Basin) accounted for 18% of China's total oil production, a substantial increase from 10% in 1995 and 13% in 2000. In particular, from 2007 onwards, the output at both Xinjiang and Changqing Oil Fields exceeded the production at the Liaohe Oil Field (Figure 4).

The majority of onshore oil and gas field exploration and development projects in China have been carried out by CNPC. China National Offshore Oil Corporation (CNOOC), founded in 1982, has been in charge of offshore oil fields. China Petrochemical Corporation (Sinopec), also founded in 1982, has been in charge of the downstream oil sector, i.e. oil refineries, marketing, and petrochemical plants. CNPC used to be placed under the supervision of the Division of Energy — the equivalent of the Ministry of Energy — which was abolished in March 1993 and then came under the direct supervision of the State Council. As part of the restructuring of the oil industry associated with the administrative reforms in 1998, CNPC exchanged some of its oil fields and refineries for those of Sinopec, whereby two major oil groups in charge of all sectors of the industry, from

[5] For example, the Daqing Oil Field, which began production in 1959 and which had maintained a capacity of over 50 million tonnes per year during the period from 1979 to 2003, produced only 40.20 million tonnes (approx. 800,000 barrels per day) in 2008. The combined output at these three top oil fields in 2008 accounted for 42% of the total oil production, a substantial decline from 68% in 1995 and 59% in 2000.

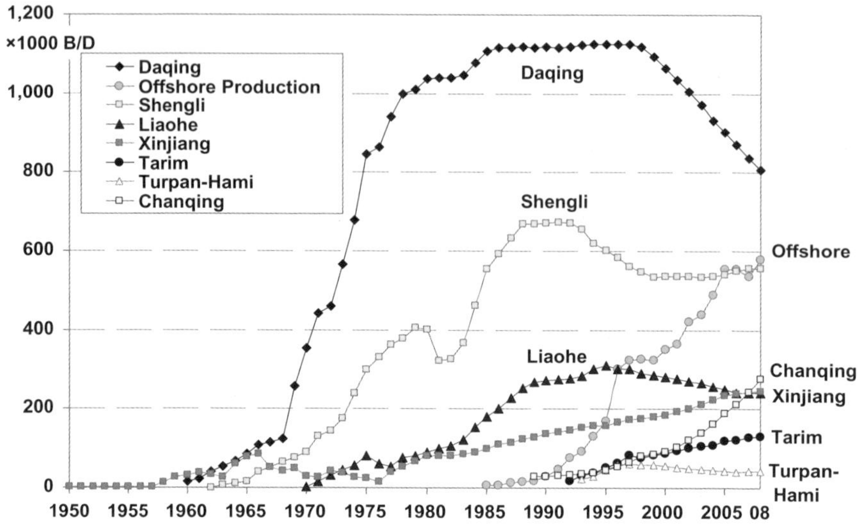

Figure 4 China's Petroleum Production from Major Oil Fields

Source: Compiled based on China's Council of International Petroleum Economics, *International Petroleum Economics Monthly*.

field exploration to refining and marketing, were created in July 1998. Through this mutual exchange of assets, Sinopec acquired the Shengli Oil Field in the Bohai Bay Basin.

CNOOC is mainly in charge of offshore crude oil and natural gas exploration and development.[6] Offshore oil output is increasing, especially in the Bohai Bay area, with the discovery of the Peng Lai Oil Field by ConocoPhillips in 1999. Thereby the share of CNOOC in China's total oil output increased from 6% in 1995 to 11% in 2000 and 15% in 2008.[7]

[6] At the moment, all of China's state-owned oil companies are able to obtain a license for both onshore and offshore exploration and development. However, against the historical background of onshore exploration and development above a depth of 5 m being allocated to CNPC and offshore exploration and development below a depth of 5 m to CNOOC, onshore development is mainly done by CNPC (PetroChina) and offshore by CNOOC.

[7] The output of CNOOC in 2008 was 195.4 million barrels of oil equivalent (boe), a 14% increase from 2007. CNOOC built the 240,000 b/d Huizhou refinery in China's southern Guangdong Province. The refinery is designed to process sour, heavy crude from

A priority mechanism has been set for drawing foreign investments into China's oil industry. The utmost priority has been given to offshore oil field exploration and development projects that have the largest amount of risk, followed by onshore oil field exploration and development projects, joint venture refineries, oil terminals, the wholesale sector, and lastly the retail sales sector of the industry. The government has granted PetroChina (a holding subsidiary of CNPC) and Sinopec the rights to conclude an onshore oil agreement with foreign companies, and granted CNOOC the rights to conclude an offshore oil agreement with foreign companies.[8]

4.3.2 Oil Imports

China imported 178.9 million tonnes (3.59 million barrels per day) of crude oil in 2008, a 9.6% increase from 163.2 million tonnes (3.28 million barrels per day) in the previous year. Together with its imports of petroleum products of 41.5 million tonnes (0.83 million barrels per day), 51% of China's oil consumption depends on imports. On the other hand, it exported 3.73 million tonnes of crude oil (75,000 barrels per day) in 2008, a 2.5% decrease from 3.83 million tonnes (77,000 barrels per day) in the previous year. As a result, the net import amount in 2008 was 198.9 million tonnes (4.00 million barrels per day), a 9.7% increase from 181.4 million tonnes (3.64 million barrels per day) in the previous year (Figure 5).

With the increase in its crude oil imports, it has become a critical issue for China to secure a stable supply of imported oil. According to the 2008 data on the crude oil import share per region, the Middle East supplied 50.1% of China's total imports, with 20.3% of the total (730,000 barrels per day) coming from Saudi Arabia. In 2005, the Middle East region

CNOOC's offshore blocks in China's northern Bohai Bay. It began commercial operations in May 2009.

[8] In October 1993, international bidding took place for five areas in the Tarim Basin in the western part of China, whereby onshore oil-prospecting areas were opened to foreign capital for the first time.

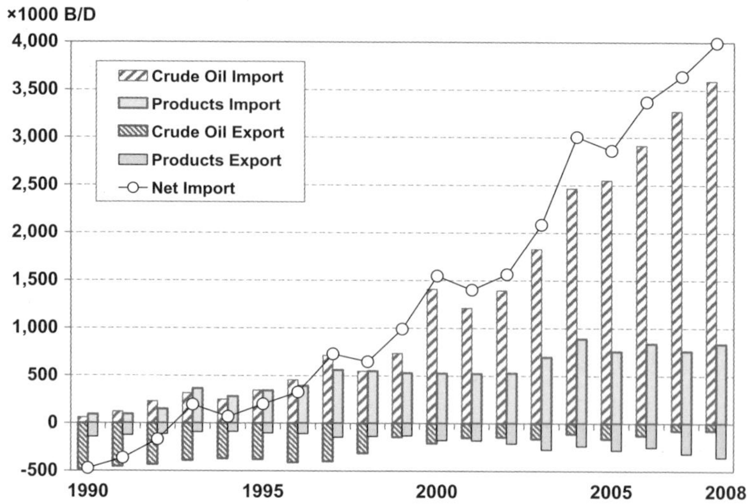

Figure 5 China's Petroleum Balance

Source: Compiled based on China's Council of International Petroleum Economics, *International Petroleum Economics Monthly*.

supplied 47.2% of the total oil imports. China is becoming more and more dependent on the Middle Eastern oil supply. Crude oil from the Middle East includes imports from Iran, which is placed under international sanctions for its nuclear programme. China's petroleum import volume from Iran reached 428,000 barrels per day in 2008, 11.9% of its total oil imports.[9]

The second-biggest source region is Africa, providing 30.2% of the total crude oil imports in 2008. The breakdown shows that Angola supplied 600,000 barrels per day, taking the second-biggest share of 16.7% after Saudi Arabia in China's total crude oil imports. Sudan, which

[9] Sinopec entered into a US$2 billion memorandum of understanding with Iran in 2004 for the development of the Yadavaran Field. Sinopec converted it into a contract in December 2007. The first-phase production rate target is 85,000 b/d by 2011, which ultimately is believed to be 300,000 b/d. First production from Yadavaran came on stream in the middle of 2009 at a rate of 20,000 b/d. In August 2009, Sinopec secured a $3 billion refining deal with the state-owned National Iranian Oil Products Distribution Company (NIOPDC) to construct a 360,000 b/d Greenfield refinery and upgrade the Abadan refinery's current 460,000 b/d nameplate capacity by 2013.

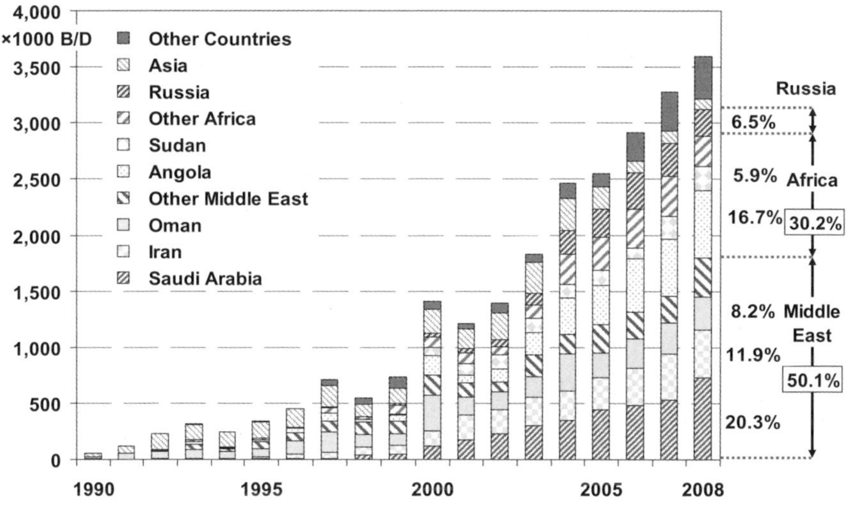

Figure 6 China's Crude Oil Imports

Source: Compiled based on China's Council of International Petroleum Economics, *International Petroleum Economics Monthly*.

provoked an international outcry over genocide in its Darfur region, also supplied 211,000 barrels of crude oil a day, 5.9% of the total imports. It was a substantial increase from 133,000 barrels a day in 2005 (Figure 6).[10]

One of the elements which have given rise to the situation described is the loss of exporting capacity by traditional major oil supply countries in Asia, such as Vietnam, Malaysia, and Indonesia, while Chinese oil imports continued rising every year. In recent years, in addition to imports from the Middle East and Africa, oil imports from Russia have been increasing. The import amount in 2008 was 234,000 barrels per day, accounting for 6.5% of the total oil imports. There used to be no pipelines between the two countries, and as such, under the supply contract entered into in January 2005 between Rosneft and CNPC, China's crude imports

[10] CNPC has a 40% equity interest in the Heglig and Unity Oil Fields in Blocks 1, 2, and 4 and in producing Nile Blend crude oil. CNPC, Sudapet (Sudan's national oil company), and other international investors have also produced crude oil from Blocks 3 and 7 since 2006. CNPC pumps oil from the Fula Oil Field as an operator and sends the oil through a pipeline to the 100,000 b/d Khartoum refinery, a 50:50 joint venture between CNPC and Sudapet.

were all delivered by railway via major rail loading facilities in two Sino-Russian towns, Zabaikalsk and Naushk.[11] In November 2007, an agreement was reached between Russia and Kazakhstan to transport 100,000 barrels of Russian crude oil per day to China via the Atasu-Alashankou pipeline. In 2008, TNK-BP and Gazprom Neft began crude shipments to China from their facility in West Siberia through this pipeline.

4.3.3 Overseas Equity Procurement

CNPC and Sinopec, both state-owned oil companies with strong support from the government, have been actively looking for overseas oil resources. In recent years, CNOOC has also taken steps to expand overseas. Some examples of their activities in recent years are as follows:

- In June 2009, Sinopec bought Swiss oil giant Addax Petroleum for C$8.27 billion (US$7.2 billion). Addax Petroleum has oil operations in West Africa and the Middle East. It produced in West Africa, including Nigeria, 134,700 barrels of oil per day during the first quarter of 2009. However, the Iraq government decided in October 2009 to strip Sinopec of the right to take part in a second round of bidding for its major oil fields, on the grounds that Addax Petroleum was developing the Taq Taq Oil Field in Iraq's autonomous Kurdish region. The central government of Iraq has branded all Kurdistan Regional Government oil contracts illegal. The large-scale acquisition strategy by Sinopec is causing some friction.
- In July 2007, together with CNOOC, Sinopec also bought a 20% stake in an Angolan offshore deepwater block from US oil major Marathon Oil. The offering price is estimated to have been $1.3 billion. CNPC also proposed in July 2009, together with CNOOC, to bid an estimated $17 billion for Spanish oil major Repsol YPF's Argentinean unit.

---

[11] The contract provided for Rosneft to supply 366.5 million barrels of crude oil to China till 2010. It was renewed in February 2009 and Rosneft expects to supply 300,000 barrels of crude oil per day to China under the new deal beginning in January 2011 for 20 years. In return, China Development Bank agreed to loan to Rosneft and pipeline company Transneft $15 billion and $10 billion, respectively, over a 20-year period.

- In May 2009, PetroChina acquired Keppel Corporation's entire 45.51% stake in Singapore Petroleum Company (SPC) for $1.02 billion, a 24% premium to SPC's closing price. PetroChina acquired more shares in SPC and boosted its stake in this Singaporean company to 70.13% in July 2009. SPC is an energy company involved in refining as well as exploration and production of petroleum in Indonesia, Vietnam, Australia, and China. Singapore Refining Company (SRC), a joint venture between SPC and Chevron and in which SPC has a 50% stake, has a total capacity of 290,000 barrels per day. It is one of the three major petroleum refiners in Singapore, the other two being ExxonMobil and Shell.
- Furthermore, in September 2009, China Investment Corp. (CIC), a sovereign wealth fund responsible for managing China's huge foreign exchange reserves, purchased 11% of the global depository receipts of Kazakhstan-based KazMunaiGas Exploration & Production (KMG EP) for $939 million. KMG EP is a subsidiary of the National Company KazMunaiGas (NC KMG) and is involved in exploration and development. The government of Kazakhstan offered 39% of KMG EP's shares in 2006. The output of KMG EP in 2008 reached 11.95 million tonnes.

4.3.4 Measures to Secure Oil and Gas Transportation Routes

As noted earlier, the Middle East and Africa accounted for 50% and 30%, respectively, of China's total oil imports in 2008. The majority of these oil imports are transported by tanker vessels from the Strait of Hormuz through the Indian Ocean via the Strait of Malacca to the Chinese coast. These sea lanes, via the politically unstable Strait of Hormuz in the Middle East and the busy Strait of Malacca, are making it more and more difficult for very large crude carriers to go through. These choke points pose an extremely high risk for China in its efforts to secure a stable crude oil supply.[12]

[12]The Strait of Malacca links the Indian Ocean to the South China Sea and Pacific Ocean. It is the shortest sea route between oil-producing countries in the Persian Gulf and oil-consuming countries in Asia. The Strait of Malacca is only 2.7 km wide at its narrowest

This is why measures are being implemented to secure onshore transportation routes in addition to offshore routes, including a shift in transportation medium of Russian crude oil from railway to pipeline and the construction of a new pipeline transportation route in Xinjiang Uyghur Autonomous Region to import Central Asian crude oil from Kazakhstan and others. In addition, in order to avoid transportation via the Strait of Malacca, consideration is being given to develop a pipeline transportation route from Myanmar to Kunming in Yunnan Province in China. In June 2009, CNPC and Myanmar's Ministry of Energy signed a memorandum of understanding on the joint development of the 771 km cross-country pipeline that will ultimately have an annual transmission capacity of 22 million tonnes (440,000 barrels per day).

As for natural gas transportation, a gas pipeline is under construction by CNPC from Turkmenistan via Uzbekistan and Kazakhstan to China. Construction in Uzbekistan and in Turkmenistan was completed in September and October 2009, respectively. The length of the pipeline is 188 km in Turkmenistan, 530 km in Uzbekistan, and 1,300 km in Kazakhstan. The pipeline has the capacity to transport up to 40 billion cubic metres of natural gas annually. The whole pipeline will be completed in December 2009, and the CNPC plans to construct a 4,500 km pipeline in China for the second stage of the project.

4.3.5 Strategic Oil Reserves

Oil stockpiles are important in order to respond to emergencies. The government plans to increase its strategic petroleum reserve by 2.6 times to 270 million barrels over five years. The construction of the four first-phase national oil stockpile terminals located in Zhejiang Province and others began in 2003 and were nearly completed in 2009. These terminals have been filled

---

point, which creates an opportunity for shore-based pirates to attack vessels. As Japan's economy heavily depends on the safe passage of vessels through the Strait, Japan has long cooperated with Singapore, Malaysia, and Indonesia in the area of navigation safety through joint research, sharing of equipment, mapping water depth charts, training, etc. Japan has also aided civilian law enforcement capabilities through its Coast Guard vessels. Japanese Coast Guard vessels have patrolled and carried out joint exercises with civilian maritime counterparts.

with 120 million barrels of crude oil.[13] The second phase of the work will commence before the end of 2009 in eight locations to construct national oil reserve terminals. Their total capacity will amount to 169 million barrels, with an estimated investment of $4.39 billion. With the completion of the first phase of construction of strategic oil reserves, China came to have a 20-day supply of crude oil stockpiles. The government plans to have enough strategic oil reserves to last 90 days, on a par with other developed countries.

4.4 TRADE WITH JAPAN

4.4.1 Resumption of Trade between Japan and China: L-T Trade

In 1949, when the People's Republic of China was founded, Japan was occupied by the Allied Forces and was under the control of the General Headquarters (GHQ). Japan was not granted the authority to decide whether to recognise the Republic of China (Taiwan) or the People's Republic of China (China) as a sovereign state. In September 1951, the San Francisco Peace Treaty between the Allied Forces and Japan was signed in San Francisco by 49 nations. At the same time, the Security Treaty between Japan and the United States was also signed, whereupon Japan built an alliance with the United States.[14]

The Korean War broke out in June 1950 and the Chinese troops intervened in October the same year. Therefore, it became impossible to have formal contact with China and only unofficial exchanges through the Japan-China Friendship Association established in 1950 remained. In December 1950, a total ban was placed on exports to China, though some representatives in the Japanese government strongly argued that trade relations with China should be maintained.[15]

---

[13] These terminals are located in Zhenhai and Zhoushan in Zhejiang Province, in Huangdao in eastern Shandong Province, and Dalian in northeastern Liaoning Province.

[14] As a result, it would not be an option for Japan to establish diplomatic relations with Communist China, i.e. People's Republic of China. The Sino-Japanese Peace Treaty was signed in 1952 between Japan and Taiwan (Republic of China).

[15] In May 1945, bodies such as the Japan-China Trade Promotion Association and the Parliamentary Leagues for the Promotion of Trade between Japan and China came into

With the ceasefire that ended the Korean War being reached in July 1953, both houses of the Diet, i.e. the House of Representatives and the House of Councillors, adopted a "resolution favouring an increase of trade with the People's Republic of China". In October the same year, the second Japan-China private trade agreement was concluded, whereby trade with China at the private-sector level commenced. However, in the wake of the Nagasaki National Flag Incident in May 1958, China strongly criticised the action taken by the Japanese government in response to the incident and trade between Japan and China was completely severed for two and a half years.[16]

On the basis of the separation of politics and economics, a theory of separating political interests from economic interests, Japan continued negotiations for reopening trade. However, China did not concede the principle of non-separation of politics and economics. After the collapse of the Kishi administration, which had practised hard-line policies towards China, the Ikeda administration took office in July 1960 and actively explored the possibilities of reopening trade between Japan and China. In response, Zhou Enlai, the Premier of China, announced the "Three Principles of Trade", which paved the way for the resumption of trade on the basis of the so-called "friendly trading companies formula".[17]

A Memorandum on Comprehensive Trade between Japan and China signed in November 1962 was a semi-private, long-term barter formula. This memorandum was signed by Tatsunosuke Takasaki of the Liberal Democratic Party, who led the Economic Mission to China, and Liao Chengzhi, Chairman of the Asia and Africa Unity Committee of China.

being. In June 1952, Japan-China Trade Promotion Association members visited Beijing. They signed the first Japan-China private trade agreement against the policy of the Japanese government, which caused much controversy.

[16] Mobs belonging to a rightwing Japanese political group stormed a Chinese products fair held in the city of Nagasaki and dragged down and caused damage to a Chinese national flag.

[17] Trade agreements on a government-to-government level (all agreements can provide security only when signed by both governments), contracts on a private basis, export of specific items by special arrangement from China to Japan.

This agreement was called "L-T Trade" after their surnames.[18] The L-T Trade Agreement also functioned, to a certain degree, as an official point of contact between Japan and China for negotiations until the normalisation of relations in 1972.

Under the L-T Trade Agreement, the scope of trade activities between the two countries expanded at once to reach a total of US$200 million in 1966, accounting for over one third of total trade between Japan and China. In 1968, one year after the expiration of the five-year L-T Trade Agreement period, the Sino-Japanese Memorandum Trade Meeting Communiqué was signed, and thereafter a memorandum was signed every year by representatives of both countries in charge of negotiations. This trade relation continued until 1973, one year after the normalisation of relations, and played a role in promoting economic exchanges between Japan and China.

4.4.2 Normalisation of Relations between Japan and China

In the late 1950s, with the campaign against Stalin by Nikita Khrushchev, Premier of the Soviet Union, relations between the Soviet Union and China deteriorated over ideological differences as well as territorial issues. In 1969, tensions between the two countries culminated in armed clashes over Damansky Island located on the Ussuri River on the border between the two countries, raising the spectre of an all-out war between China and the Soviet Union. Mao Zedong, Chairman of the Communist Party of China, ready to confront the Soviet Union, approached the United States for a closer partnership. This resulted in the visit of US President Richard Nixon in 1972 to China as a step in improving relations between the United States and China.[19]

---

[18] Japan would export chemical fertilisers, steel, pesticides, agricultural machinery, plants, etc. China would export coal, iron ore, maize, beans, etc. The period from 1963 to 1967 was the first five-year trade period, aimed at an average annual trade of £36 million. Although there was no formal diplomatic relation between the two countries, liaison offices were set up, through which semi-private trade was carried out with government guaranteed loans. Large-scale plants, such as Baoshan Iron & Steel and Daqing Petrochemical Company, were granted long-time credit under this L-T Trade Agreement.

[19] US President Richard Nixon visited China in February 1972 at the invitation of Premier Zhou Enlai of the People's Republic of China. He met with Chairman Mao Zedong of the

In these circumstances, the question of normalisation of Sino-Japanese relations became a major issue for general elections within the Liberal Democratic Party, pro-American conservative administration. In 1972, Kakuei Tanaka became Prime Minister on a manifesto promising the normalisation of relations between Japan and China.[20] In September of the same year, the Joint Communiqué of the Government of Japan and the Government of the People's Republic of China was signed in Beijing between him and Zhou Enlai, Premier of the State Council.

The Communiqué stated,

> Japan and China are neighbouring countries, separated only by a strip of water with a long history of traditional friendship. In spite of the differences in their social systems existing between the two countries, the two countries should, and can, establish relations of peace and friendship. The normalisation of relations and development of good-neighbourly and friendly relations between the two countries are in the interests of the two peoples and will contribute to the relaxation of tension in Asia and peace in the world.

It also confirmed nine issues, including the following:

- The abnormal state of affairs that has hitherto existed between Japan and the People's Republic of China is terminated on the date on which this Joint Communiqué is issued.
- The government of Japan recognises the government of the People's Republic of China as the sole legal government of China.
- The government of the People's Republic of China reiterates that Taiwan is an inalienable part of the territory of the People's Republic of China. The government of Japan fully understands and respects this stand of the government of the People's Republic of China, and it firmly maintains its stand under Article 8 of the Potsdam Proclamation.

The Chinese side argued that Taiwan was a part of China and claimed the right to annex Taiwan. The Japanese government did not agree with

---

Communist Party of China and the Joint Communiqué of the United States of America and the People's Republic of China was issued in Shanghai.

[20] His term of office was from July 1972 to December 1974.

this Chinese position and used the wording "understand and respect" instead of "recognise",

Thereafter, the Trade Agreement between the government of Japan and the government of the People's Republic of China took effect in June 1974. In August 1978, in light of the Joint Communiqué of the Government of Japan and the Government of the People's Republic of China, the Treaty of Peace and Friendship between Japan and the People's Republic of China was concluded.

4.4.3 Progress in Trade between Japan and China

The total amount of trade between Japan and China was US$1.1 billion in 1972, when the relations between the two countries were normalised. After 1978, when China adopted the reform and opening-up policies, the scale of trade between Japan and China surged to reach over US$10 billion in 1981. As a result, each came to assume a very important position in the trade and economy of the other, thereby further deepening economic interdependency.

The trade surplus between Japan and China continued for Japan until the mid-1980s. However, from the late 1980s onwards Japan ran a trade deficit with China. Since 2001, when China became a member of the World Trade Organization, the share of China in Japan's total trade has surged. It exceeded the amount of trade between Japan and the United States in 2007. One major factor which underlay the expansion of trade between Japan and China was the fact that Japanese companies had shifted their production bases to China. These accelerated trends that China would import core parts it could not procure domestically from Japan while Japan would import from China final products. Exports of products to Japan by foreign companies based in China and Chinese companies are also increasing. The trade between Japan and China rose to US$266.3 billion in 2008 (up 12.5% from the previous year), hitting a record high for 10 straight years and exceeding the trade between Japan and the United States for two consecutive years. Of the US$266.3 billion in 2008, exports and imports accounted for US$124 billion (up 13.7%) and US$142.3 billion (up 11.5%), respectively (Figure 7).

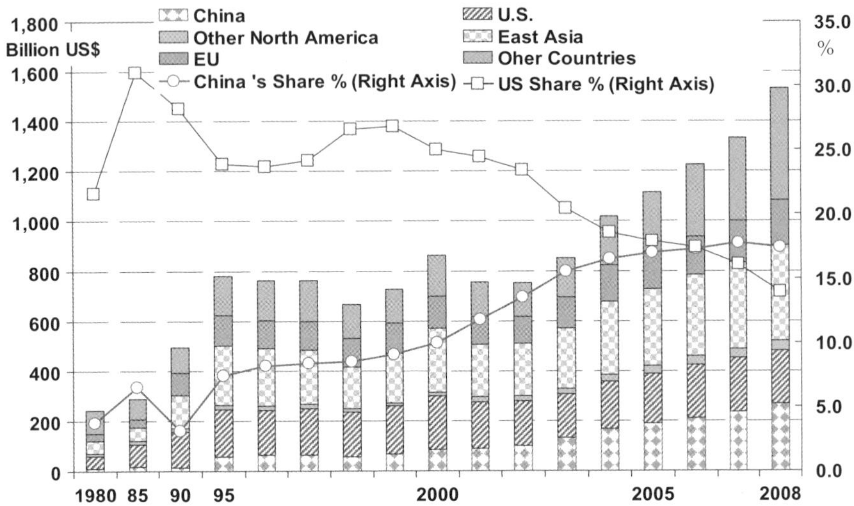

Figure 7 Japan's Foreign Trade

Source: Japan External Trade Organization (JETRO).

On the other hand, the share of Japan-China trade in China's total trade was the highest in 1978 but dropped in the 1980s. From 1992 onwards, Japan's share had increased again, only to be back on a downward track in August 2001 when Prime Minister Koizumi visited the Yasukuni Shrine and invited a strong backlash from China. Japan's share has declined since then. In 2002, trade with the United States exceeded trade with Japan. In 2008, Japan accounted for 9.2% of China's total trade, while the share of the United States reached 12.9% (Figure 8).

4.5 ENERGY TRADE BETWEEN JAPAN AND CHINA

4.5.1 Oil Trade

The supply of Daqing crude oil to Japan started in April 1973 when the first oil crisis broke out. Since 1978, when the Treaty of Peace and Friendship between Japan and the People's Republic of China was concluded, the export of Daqing crude oil to Japan played the role of a so-called facilitator of trade between the two countries. China wanted to obtain enough foreign currency, which was necessary for its economic

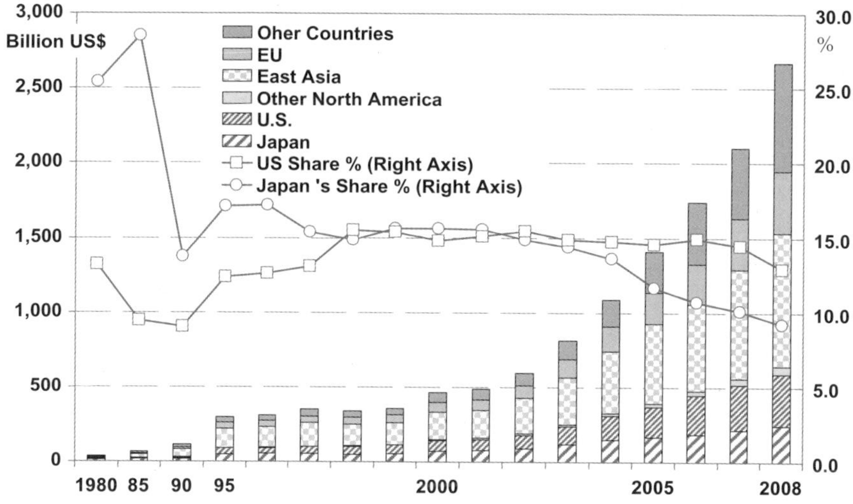

Figure 8 China's Foreign Trade

Source: JETRO.

development but had been in short supply since the Sino-Soviet dispute, by exporting its abundant energy resources to Japan.

Although China became a net importer of oil in 1993, China exported low-sulphur and high-priced Daqing crude oil to get foreign currency, while importing high-sulphur and less expensive Middle East crude oil for processing. The agreement with Japan for the period from 1996 to 2000 provided that China should export 6–8.5 million tonnes of crude oil. Accordingly, China exported 8 million tonnes of oil in 1997 and 6 million tonnes in 1998. However, Daqing crude oil production diminished after the peak, and domestic oil prices in China became higher than oil export prices. China's export volume in 1998 was no more than the lowest limit specified in the agreement and supply has been interrupted since January 2004 (Figure 9).

On the other hand, in the wake of the first oil crisis in 1973, having been heavily dependent on the Middle East, Japan diversified and extended oil supply sources to regions other than the Middle East in an effort to ensure a stable oil supply. The normalisation of relations between Japan and China in 1972 provided the best opportunity for Japan to increase oil imports from China. The amount of oil imported from China

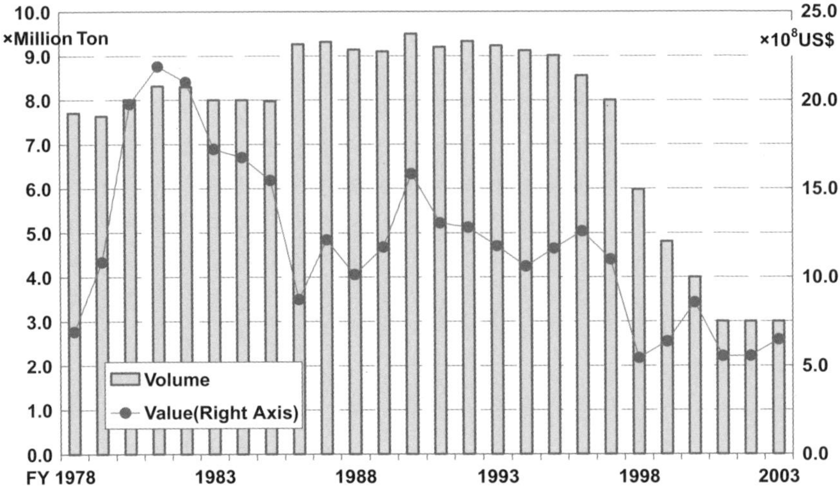

Figure 9 Japan's Crude Oil Imports from China through L-T Trade Agreement

Source: JETRO.

was 23,000 barrels per day at the beginning of 1973. Two years later, in 1975, the amount reached 163,000 barrels per day, and China's share in Japan's total oil imports increased to 3.6%.

At the same time, Japan tried to switch from oil to other energy resources, such as coal, natural gas, and nuclear, whereby its oil imports started to decline after the second oil crisis in 1980. However, Japan maintained a certain volume of crude oil imports from China. Therefore, China's share reached 7.9% in 1987. Subsequently, the oil demand in Japan once again started to grow, while China's oil export capacity continued to decline as stated before. Accordingly, China's share continued to decline to 7,000 barrels per day in 2004, i.e. virtually 0% in terms of the share in Japan's total oil imports, a year after the suspension of imports under the L-T Trade Agreement (Figure 10).

In addition to the imports of Daqing crude oil under the L-T Trade Agreement, supply was also made from Bohai Light Crude Oil in Bohai Bay operated by Japanese oil development companies, Nanhai Medium and Nanhai Blend, and Lu Feng in Pearl River Mouth in southern China. Imports of oil from these fields, however, became nil when the projects came to an end. Currently, Chinese oil is being supplied only on a

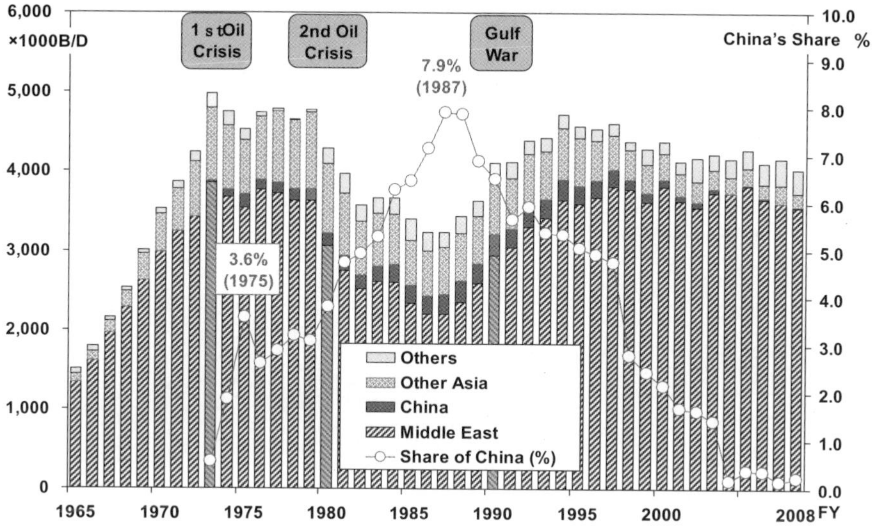

Figure 10 Japan's Crude Oil Imports

Source: METI, *Yearbook of Mineral Resources and Petroleum Products Statistics*.

commercial basis from Panyu (Hong Kong), Xijang (Hong Kong), CFD (Bohai Bay), and Peng Lai (Bohai Bay) Oil Fields operated by CNOOC and oil development companies with foreign capital.

4.5.2 Coal Trade

China has a vast amount of coal resources and produced 47% of the world coal production in 2008, i.e. 2,761 million tonnes out of 5,845 million tonnes. It is the biggest coal producer in the world. Coal accounts for approximately 73% of China's primary energy supply, and 2,760 million tonnes of coal were consumed in 2008. In China, coal is mainly used for electricity generation. About 80% of the total amount of electricity was generated at coal-fired power plants in 2007 (Figure 11). Therefore, the domestic demand for coal is increasing along with an increase in domestic demand for electricity. Most of the coal mines are located in inland China, e.g. Shanxi, Shaanxi, and Inner Mongolia. On the other hand, power stations are located closer to the market of electricity along the coast. Coal is transported to these power stations mainly by railway.

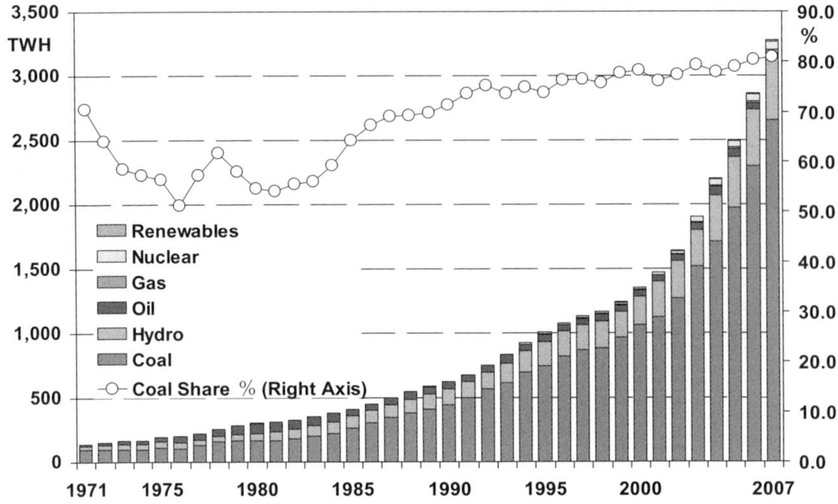

Figure 11 China's Electric Power Generation by Type of Energy

Source: IEA, *Energy Balances of Non-OECD Countries.*

Consequently, the expansion of railway transportation capacity has also become an issue.[21]

Coal exports have been a major contributor for China in getting foreign currency. However, its export capacity started to decline after the peak in 2003. Giving priority to fulfilling the domestic demand for coal, the Chinese government introduced a coal export quota. In 2004, the export licensing system was implemented to regulate its annual total coal export volume. China's coal export volume in 2008 was 47.4 million tonnes, a decrease of 6.3 million tonnes from 53.7 million tonnes in the previous year.[22]

On the other hand, coal imports from overseas by coal-fired power plants along the coast in southern China had increased sharply to reach 45.6 million tonnes in 2008. Against this background, coal exports to Japan under the L-T Trade Agreement also dropped after the peak of

[21] Coal accounted for approximately 49% of China's total volume of railway cargo transportation, i.e. 1.544 billion tonnes out of 3.142 billion tonnes in 2007.

[22] The maximum quota was 80 million tonnes in 2006, which dropped to 70 million tonnes in 2007 and 53 million tonnes in 2008.

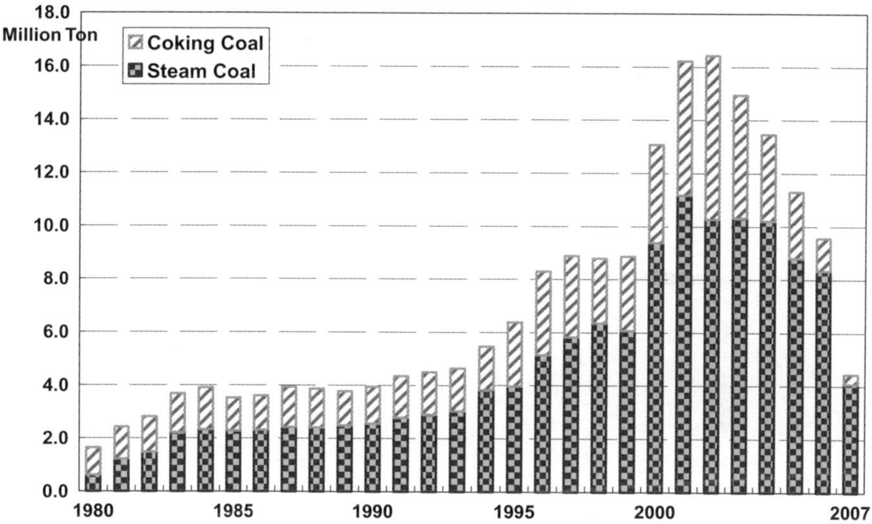

Figure 12 China's Coal Exports to Japan through L-T Trade Agreement
Source: JETRO.

16.4 million tonnes in fiscal year 2002 to 4.4 million tonnes in fiscal year 2007 (Figure 12).[23]

Combining the volume of coal imports through general transactions with that of coal imports under the L-T Trade Agreement, Japan's total coal imports from China reached their peak of 28.2 million tonnes in 2003, i.e. 17.4% of Japan's total coal imports. Japan's total coal imports from China fell only 5.6% to 10.3 million tonnes in 2008 (Figure 13).

4.6 OIL EXPLORATION COOPERATION BETWEEN JAPAN AND CHINA

4.6.1 Exploration and Development of Bohai Bay Oil Field

In December 1978, the 11th Central Committee of the Communist Party of China held its third plenary session. The meeting decided to stop taking

[23] The volume is on a fiscal year basis, i.e. from April to March. The contracted volume was 4.91 million tonnes for the fiscal year 2007, 5.21 million tonnes for the fiscal year 2008, and 1.33 million tonnes for the fiscal year 2009.

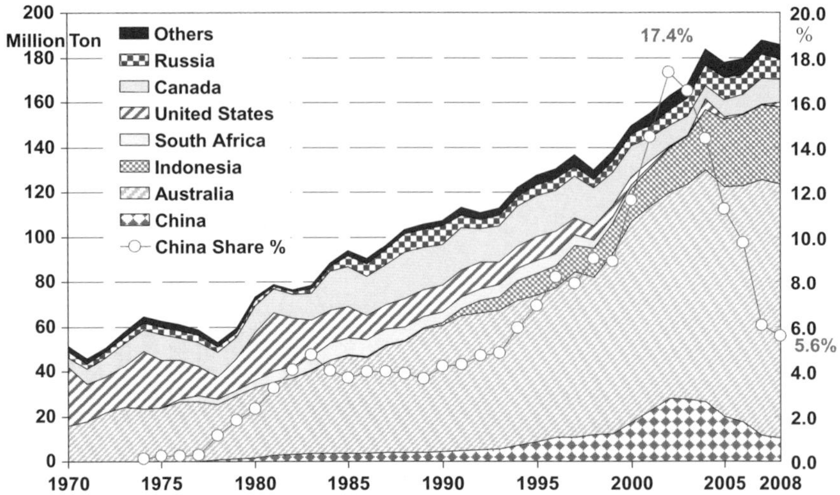

Figure 13 Japan's Coal Imports by Country

Source: Ministry of Finance, Trade Statistics of Japan.

class struggle as the key policy and shift the working focus to reform and opening-up, to meet the needs of modernisation. However, economic development was indispensable for China in order to promote its modernisation. Therefore, it was necessary to actively promote the open-door policy to bring in advanced technology, production equipment, capital, and management methods from overseas. Simultaneously, it became necessary for Japan to diversify its energy supply sources to ensure a stable supply of energy.

In July 1978, when the delegation of Japan National Oil Corporation (JNOC) visited China, cooperation between Japan and China was agreed upon as the first stage in opening oil-prospecting areas in China to overseas capital. They agreed to cooperatively explore and develop areas in the south of Bohai Bay, in Western China, and Chengbei Oil Field.[24] In May 1979, a memorandum on a line of credit of JPY 420 billion (US$2 billion) was signed between Export-Import Bank of Japan and Bank of China, which stated that necessary capital would be provided for projects

[24] In December 1979 and February 1980, JNOC (at the time) and the Offshore Branch of Petroleum Company of the People's Republic of China signed a basic agreement.

implemented by China to develop onshore oil fields on the coast of Bohai Bay (Huabei Oil Field and Shengli Oil Field) and develop coal mines at Baodian in Shandong Province.

In order to promote oil exploration and development projects at the offshore Bohai Bay as Japan's national project led by the Japanese government, two companies — Japan-China Oil Development Corporation (JCO) and Chengbei Oil Development Corporation — were set up in April 1980 with investment from JNOC and 47 private companies. These two companies were funded with 100% risk guarantee capital from the Japanese stakeholders. It was agreed that the necessary costs of $100–200 million would be invested in exploration works, and upon the success of these works, development works (at an estimated cost of $1 billion) would be carried out jointly by Japan and China.[25] From 1980 onwards, seismic surveys of areas covering 25,000 km^2 were conducted, 38 exploration wells were drilled, and 25 wells were successfully found, thereby resulting in the discovery of 11 oil prospects.

Development work was started on three oil fields, i.e. Chengbei, BZ28, and BZ34 Oil Fields. However, as the oil layers are segmented with faults, the development of these fields proved to be very difficult.[26] Originally, the two oil companies — JCO and Chengbei Oil Development — were required to keep separate accounting records, but the latter was obliged to merge with JCO in 1989. However, due to falling crude oil prices and an overvalued yen, the business performance of JCO had also worsened so

[25] At this stage of development, Japan and China would bear 49% and 51% of the capital, respectively. With regard to the capital injection by China, a loan of JPY 105 billion (approximately US$460 million at the exchange rate fixed according to US$1.00 = JPY 230.0) was provided to China from Export-Import Bank of Japan in accordance with the memorandum signed in May 1979. It was planned that, during a period of 15 years when commercial oil production started, 57.5% of crude oil would be allocated to China and 42.5% be supplied to Japan, and the capital injected by Japan would be recovered over 15 years from the proceeds of the sale of its 42.5% of the oil.

[26] Production at Chengbei Oil Field began in September 1985, BZ28 Oil Field in July 1989, and BZ34 Oil Field in July 1990. At the peak of oil production, the daily output from BZ28 Oil Field and BZ34 Oil Field was about 5,200 and 5,800 barrels, respectively, and Chengbei Oil Field produced some 5,600 barrels of oil per day. However, due to the complicated structure of the oil seams, the volume of recoverable reserves turned out to be less than originally estimated. BZ28 Oil Field stopped production at the end of September 1994.

much that it eventually posted an estimated annual loss of over JPY 1 billion.[27] In June 2000, JCO agreed with the Chinese side that it would withdraw itself from the project at the end of September of the same year at the expiry of the oil contract for the Chengbei Oil Field.

In October 2000, JNOC announced that it would dissolve JCO. With its debts exceeding assets, JCO approved its dissolution and submitted applications for commencement of special liquidation proceedings in the Tokyo District Court.[28] JNOC posted losses totalling JPY 139.2 billion in loans and investments, and the invested capital of approximately JPY 50 billion from 45 private companies became unrecoverable. At the time of JCO liquidation, JNOC posted losses totalling JPY 156.4 billion including a loss of JPY 38.7 billion from amortisation of claims for indemnity raised during the course of providing other debt guarantees.[29]

However, when CNOOC formed a partnership with ConocoPhillips and redirected its target of exploration from the 2,000 m depth layer below the seabed to the upper depth layer of 1,000 m below the seabed, it discovered oil reserves in 1999. Thereafter, Shell joined forces with them in 2002, and oil reserves were discovered one after another in the areas abandoned by Japan. This made it possible for CNOOC to establish a solid revenue base.[30]

---

[27] The new Japan-China Offshore Oil Company was established in September 1992 and started new exploration activities, including two exploration well drillings. The work, however, did not lead to any commercial discovery. The company was wound up in October 1995.

[28] JCO was established with JPY 100 billion in capital, of which JPY 64.5 billion was injected by JNOC and the remaining JPY 35.5 billion by private companies. The amount of accumulated deficit was JPY 50.4 billion. It had borrowings of JPY 74.1 billion (of which JPY 39.7 billion had been borrowed from JNOC) and owed JPY 45.6 billion in unpaid interest. As borrowings of JPY 13.4 billion from JNOC were waived, its actual deficit was a total of JPY 109.4 billion (JPY 50.4 billion + JPY 45.6 billion + JPY 13.4 billion) and exceeded its assets.

[29] JNOC's loss was calculated at JPY 156.4 billion in total, with its breakdown being JPY 64.5 billion in capital invested in JCO, a loan of JPY 53.1 billion (JPY 39.7 billion + waived JPY 13.4 billion), and JPY 38.7 billion in debt guarantee for borrowings from other banks. This excludes JPY 34.7 billion in unpaid interest.

[30] CNOOC submitted a bid in June 2005 to take over US Unocal at US$67 per share, totalling US$18.5 billion. The bid, however, met with political opposition from the US political circle which viewed the proposed merger as a threat to the US. In August 2005, CNOOC announced that it had withdrawn its offer for Unocal.

Apart from the activities indicated, Japanese oil companies also carried out other exploration and production projects in China, which are listed here. All these projects have already been terminated. Currently, no exploration or production activity is being carried out by Japan in China.

- **Oil development in the South China Sea:** In the second round of bidding in 1984 for areas in the South China Sea, Block 16/06 located in the Pearl River Mouth Basin was granted to Japan. New Nanhai Oil Development Co., Ltd., New Huainan Oil Development Co., Ltd., and NMC Pearl River Mouth Oil Development Co., Ltd., jointly set up JHN Oil Operating Company. JHN discovered in February 1987 the Lu Feng 13–1 Oil Field which came into operation in November 1993. At the peak of oil production, 25,000 barrels of crude oil were produced daily from this oil field. Its accumulated crude oil production till February 2009, when the production period under the contract came to an end, was approximately 70 million barrels. At the expiry of the production period, the project was wound up, and the operation was transferred to the Chinese side.
- **Oil development in the Tarim Basin:** In the first round of bidding in 1993 for Chinese onshore oil-prospecting areas, Japanese companies secured the license to Block 1 in the Tarim Basin jointly with Italian company Agip. Upon conclusion of a contract with CNPC in February 1994, the exploration work began in May 1994. Following another round of seismic surveys, exploration well drillings were conducted, which did not result in the discovery of oil. The area was abandoned in April 1998.

  Japanese companies together with US ExxonMobil also won the bid for Block 3 located in the southeastern part of the Tarim Basin and carried out exploration and development activities. However, no oil was discovered and they withdrew from the project in March 1998.
- **Oil development in the East China Sea:** In the fourth round of bidding in December 1993 for offshore oil-prospecting areas, Teikoku Oil Co., Ltd., and Japan Petroleum Exploration Co., Ltd., won Blocks 41/17 and 42/03 in the East China Sea. In February 1995, exploration well drillings were carried out for one well on each of these blocks, but no oil was discovered.

4.7 CONFLICT OF INTEREST BETWEEN JAPAN AND CHINA OVER MEASURES TO SECURE ENERGY SUPPLY

4.7.1 Eastern Siberia–Pacific Ocean (ESPO) Pipeline

In 1998, a specific plan was drawn up between Yukos (Russia), Transneft (Russia), and CNPC (China) to construct a pipeline to transport West Siberian oil from Angarsk in Russia to Daqing in China (2,260 km long Daqing route). In February 1999, a framework agreement was signed to investigate its commercialisation. The plan was agreed upon in July 2001 between the Chinese and the Russian governments.

However, in January 2002, as an alternative to the proposal from Yukos, Transneft submitted to the government a plan to construct a crude oil pipeline leading into the Sea of Japan from Angarsk to the port in Perevoznaya in the Hasan district in the Maritime region.[31] The proposal from Transneft received strong support from the Russian government as it entailed the possibility of long-term economic development as well as the opportunity to enhance economic relations with the Asia-Pacific region. In consequence, a contract for the construction of a pipeline bound for China was not agreed at the Sino-Russia Summit in December 2002.

On the other hand, when Prime Minister Koizumi visited Russia in January 2003, an idea was presented to construct a pipeline from Angarsk to Nakhodka (4,000 km long Nakhodka route), and Transneft showed its strong interest therein. The Japan-Russia Action Plan was agreed upon by Japan and Russia, whereby the Japanese government pledged its support and financial assistance to resource development in Sakhalin and Siberia and for the construction of the pipeline. Therefore, it became unlikely that the Yukos plan would be adopted.[32]

---

[31] For Russia, the Daqing route had certain benefits in that its share of construction costs would be smaller. However, as the destination of this route would be limited to China, Russia was concerned that it might thus lose control over the price.

[32] Subsequently, Mikhail Borisovich Khodorkovsky, Yukos's founder and CEO (at the time), was arrested in October 2003 on charges of corporate and income tax evasion. In December 2004, Yukos was bought by state-owned Rosneft.

The Russian government announced in May 2003 the construction of a crude oil pipeline with a feeder line leading to Daqing. In 2004, the site for the starting point was changed from Angarsk to Taishet. The Russian government decided in April 2005 to construct the 4,700 km long pipeline in two phases: in inland Russia during Phase 1 and from inland Russia to the Pacific Ocean during Phase 2. In the initial phase, a 2,400 km long pipeline from Taishet to Skovorodino would be laid (with the capacity to transport 600,000 barrels per day), while Phase 2 entails a line extending further from Skovorodino to Kozmino at Perevoznaya Bay in the vicinity of the Pacific Ocean port of Nakhodka.[33]

Subsequently, the relationship between Russia and China improved and, with the conclusion of a long-term oil supply contract between Rosneft and CNPC in January 2005, Russia decided to prioritise the construction of the line leading to China.[34] In October 2006, China and Russia began a feasibility study on the construction of a proposed oil pipeline spur from Skovorodino to the Chinese border. Transneft started in April 2009 the construction of the 64 km long feeder line leading to the border with China, which was completed in October 2009. It has the capacity to deliver 15 million tonnes per year (300,000 barrels per day). The delivery will start in October 2010 when the construction of the Chinese section of the feeder line is completed. On the other side, the construction of the main line leading to the Pacific Ocean has been delayed due to a subsequent sharp price rise of materials, and the work is expected to be completed around 2012 at the earliest. Until then, 300,000 barrels of oil per day will be transported by railway from Skovorodino to Kozmino, and it is expected that exports from Kozmino will start in December 2009[35] (Figure 14).

---

[33] In September 2006, the north route away from Lake Baikal was adopted, and the construction of the pipeline from Taishet started in April 2006.

[34] A shortage of crude oil to be transported is also pointed out. While the pipeline has the capacity to deliver 80 million tonnes per year, only 30 million tonnes of crude oil are available for transportation. Therefore, there may be certain reasons that oblige Russia to give priority to the transportation for China. The Chinese section of this feeder line is 965 km long.

[35] In October 2009, 72 railway tank wagons, each loaded with 60 tonnes of oil, arrived in Kozmino.

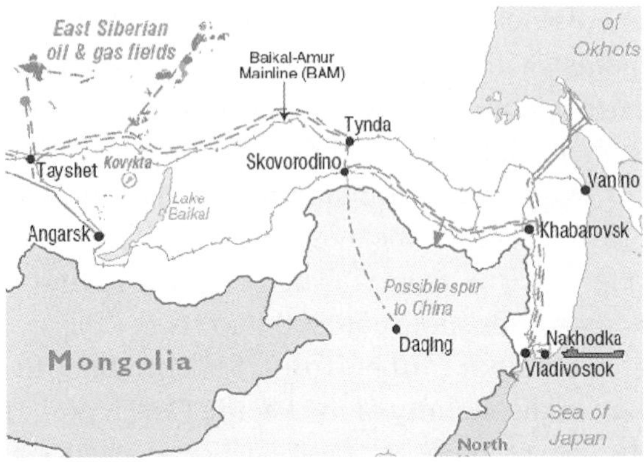

Figure 14 Eastern Siberia–Pacific Ocean (ESPO) Pipeline

Source: U.S. DOE/EIA, Country Analysis Brief, Russia, http://www.eia.doe.gov/cabs/Russia/Full.html.

4.7.2 Development of Gas Fields in the East China Sea

In June 2006, under the administration of Prime Minister Koizumi (who was in office from April 2001 to September 2006), the issue concerning development of gas fields in the East China Sea surfaced when it was confirmed that China had started development work in the area close to the "median line between Japan and China". The line divides the Exclusive Economic Zone (EEZ) and the continental shelf in the East China Sea between Japan and China.[36] During the conference of the Japanese and Chinese foreign ministers in June 2004, Japan asked China to disclose details of China's exploration data such as demarcation of the gas-prospecting areas on the Chunxiao Oil and Gas Fields. In response, the Chinese side avoided making specific reference to the submission of data and consistently maintained that it would not accept the median line that Japan unilaterally regards as the sea boundary between the two sides.

[36] The Chunxiao Oil and Gas Fields are located in the southeast of Shanghai. In July 1995, the first exploration well drilling ended successfully. They are comprised of four gas fields, i.e. Chunxiao (known as Shirakaba in Japan), Tianwaitian (known as Kashi in Japan), Longjing (known as Asunaro in Japan), and Duanqiao (known as Kusunoki in Japan). A comprehensive plan to develop the Chunxiao Oil and Gas Fields was approved by the State Planning Commission and officially announced in March 2002.

In April 2005, Japan published the result of its own investigation that the geological structures of the Shirakaba (Chunxiao in China), Asunaro, and Kusunoki fields extended into the Chinese side. In July of the same year, the right was granted to Teikoku Oil Co., Ltd., to carry out exploration well drillings in gas fields on the Japanese side of the East China Sea. On the other hand, China also started, in September 2005, drillings at Tianwaitian (known as Kashi in Japan) Gas Field. In August 2006, Koizumi paid his sixth visit to the Yasukuni Shrine on the anniversary of Japan's Second World War surrender, thereby resulting in rapid deterioration of the relationship between Japan and China.[37]

Prime Minister Abe (who was in office from September 2006 to September 2007), having taken over from Prime Minister Koizumi, in an effort to rebuild relations with China, chose China as the destination of his first official overseas trip in October 2006 and met with Chinese President Hu Jintao and Chinese Premier Wen Jiabao. Both sides agreed to make efforts to build a "mutually beneficial relationship based on common strategic interests" and cooperate in various fields, such as energy, trade, and global warming. With regard to the dispute between Japan and China over exploration of gas fields in the East China Sea, a joint statement was subsequently issued:

> Both sides reaffirmed that, in order to make the East China Sea a "Sea of Peace, Cooperation and Friendship", both sides should firmly maintain dialogue and consultation, and resolve appropriately difference of opinions. Both sides confirmed that they would accelerate the process of consultation on the issue of the East China Sea, adhere to the broad direction of joint development and seek for a resolution acceptable for the both sides.

[37]Given that Japan had stepped up its efforts to become a permanent member of the United Nations Security Council, and a school textbook on Japan's history for junior high schools and high schools, written by the New History Textbook Movement, which includes views that can be interpreted as glamorising the history of the Japanese colonial occupation of Asia, had been approved by the Ministry of Education, Culture, Sports, Science and Technology, anti-Japanese protests erupted in April 2005 in various parts of China with the slogan "Patriotism is Innocent". A boycott of Japanese goods followed, and protesters threw stones at Japan's Embassy.

In May 2008, Chinese President Hu Jintao visited Japan and met with Prime Minister Fukuda (who was in office from September 2007 to August 2008). The joint statement between the two governments on comprehensive promotion of a Mutually Beneficial Relationship Based on Common Strategic Interests was issued. It was resolved that both Japan and China would actively engage in negotiations to set out a new global framework to tackle global warming from 2013, which is after the end of the first commitment period of the Kyoto Protocol in 2012, and promote cooperation in developing energy-saving and other technologies as well as in the transfer of such technologies.

Following this joint statement, it was agreed in June of the same year that, with regard to the East China Sea gas dispute, a joint development area should be established by the Japanese and Chinese governments in the area covering the median line.[38] However, with regard to the Kashi Gas Field (known as the Tianwaitian Gas Field in China), Japan demanded that China end its development, while China maintained its right of unilateral development. Consequently, the question of joint development of this field was left unresolved.

On 21 September 2009 Prime Minister Yukio Hatoyama, who was in New York to attend the United Nations Summit on Climate Change, met with Chinese President Hu Jintao.[39] At the meeting, touching on the

[38] As for the joint development of the Shirakaba Gas Field, Thailand-Malaysia joint development of oil reserves in the Gulf of Thailand, and Australia–East Timor sharing of maritime oil resources are good examples of the challenge. In 1990, the Malaysia-Thailand Joint Authority (MTJA) was established as a statutory body under the laws of Malaysia and Thailand in 1991 to assume all rights and responsibilities on behalf of the two governments to explore and exploit natural resources in the overlapping area known as the Joint Development Area (JDA). In January 2006, Australia and East Timor brought into force the Treaty on Certain Maritime Arrangements in the Timor Sea (CMATS Treaty) and agreed in March 2003 and brought into force in February 2007 the International Unitization Agreement for Greater Sunrise (IUA). These treaties establish a framework for the exploitation and the equal sharing of the Greater Sunrise gas and oil resources.

[39] Prior to this meeting, a delegation of the Japan-China Economic Association, including Japan Business Federation Chairman Fujio Mitarai, met with Premier Wen Jiabao on 9 September 2009. It was agreed to further promote economic cooperation between Japan and China in the fields of environmental protection and energy-saving technologies. However, with regard to environmental technology transfer to China, Japanese companies

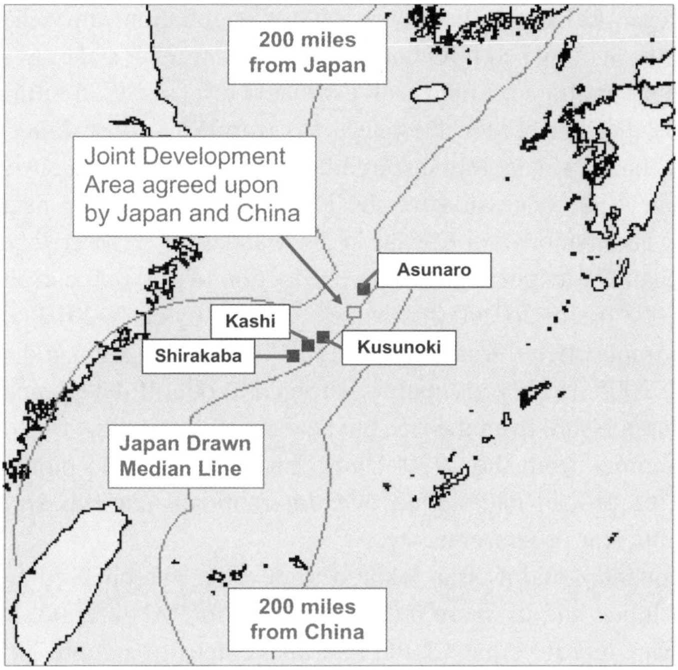

Figure 15 Disputed Oil and Gas Fields in the East China Sea

Source: The map was made by author based on information from Japanese Government (METI, Ministry of Economy, Trade and Industry).

dispute over the Shirakaba Gas Field (known as the Chunxiao Gas Field in China), which China had agreed to develop jointly with Japan, Prime Minister Hatoyama expressed concern over actions taken by China that could be seen as preparations to develop this gas field. Both sides agreed to proceed with preparations at the working level to forge a treaty on the agreement on joint development of gas fields in the East China Sea (Figure 15).

4.7.3 Oil Development in Iran

In February 2004, Japan's government-backed Inpex Corporation signed a US$2 billion deal with the National Iranian Oil Company (NIO) to develop

---

expressed strong concern about possible technology outflow and pointed out once again problems associated with intellectual property protection in China.

the Azadegan Oil Field after four years of negotiation. Inpex has a 75% stake, while state-owned NIO holds the remaining 25% stake. Azadegan is Iran's largest onshore oil field with estimated reserves of 26 billion barrels located on the Iraq border. Despite a US warning against doing business with Iran, Japan was largely inspired by the failure of Japan's Arabian Oil Co. to renew its concession for the Khafji Oil Field in the neutral zone between Saudi Arabia and Kuwait in 2000 and 2002, respectively.

Azadegan was expected to begin production of 150,000 barrels per day by mid-2008, rising to 260,000 barrels per day by early 2012, and Japan hoped to import two thirds of the output. The work had been due to start by March 2005. However, Inpex was forced to put off development of the field under pressure from the US, but also because the site was still littered with landmines from the 1980–1988 Iran-Iraq War. And more recently, talks on the project had stalled over international tensions surrounding Tehran's nuclear programme.

Consequently, and as Iran wanted to develop this oil field as soon as possible, Inpex had its share reduced by Iran to 10% in 2006. In August 2009, it was reported that CNPC had taken steps to acquire 70% of the 90% share held by the Iranian company. CNPC is said to have agreed to cover 90% of the development costs of $2.5 billion.

4.7.4 Energy Cooperation between Japan and China

China suffered serious shortages of electricity in 2004 and certain parts of southern China such as the province of Guangdong, one of the country's most heavily industrialised regions, have continued to face a shortage of electricity supply since 2008. The installed power capacity of China reached 710 million kilowatts (kW) at the end of 2007, 1.6 times the maximum demand of electric power. Therefore, its generating capacity itself is not insufficient. The shortages of electricity were due to deteriorating profits from the power generation business and a shortage of electricity transmission capacity. While all power generation costs surged with the liberalization of coal price in 2006, the government imposed price restraints on electricity. Therefore, the profit margin of electricity companies has stayed low, and the operating rates of power plants have gone down. The overall impact was that industry consumers decided to ensure their own supply through the use of diesel generators and thus further boosted domestic oil and oil product demand.[40]

In October 2005, the Chinese government led by Premier Wen Jiabao advocated in the proposal on the Eleventh Five-Year Program for National Economic and Social Development until 2010 a switch to a domestic demand–led economy and implementation of energy-saving measures to "change development patterns". With regard to energy saving, China pledged to cut energy consumption per unit of gross domestic product (GDP) by 20% by 2010, and a 10.1% reduction was achieved by 2008.

Against this background, with a view to facilitating energy-saving behaviour among the partners, the Asia-Pacific Partnership on Clean Development and Climate (APP) was set up in July 2005 with seven partner countries, i.e. Japan, China, the United States, Australia, Canada, South Korea, and India. More than 50% of global CO_2 emissions come from these seven countries. In order to reduce greenhouse gas emissions more effectively by facilitating the development, deployment, and transfer of cleaner and more efficient technologies, the partners have approved eight public-private sector task forces, covering aluminium, buildings and appliances, cement, cleaner fossil energy, coal mining, power generation and transmission, renewable energy and distributed generation, and steel, thereby promoting the transfer of energy-saving technologies. In the steel sector, some positive results have been achieved, such as the installation of coke dry quenching equipment (CDQ) at ironworks in China.[41]

[40] China's electricity output in 2008 was 3.40 TWh, a 5.5% increase from 2007. However, electricity companies suffered a loss of RMB 70 billion (US$10.3 billion). Therefore, the Chinese government put in RMB 10 billion (US$1.46 billion) to bail out five electricity companies and two electric power transmission companies.

[41] CDQ is the equipment which quenches red-hot coke in the cooling process by means of circulating inert gas in a quenching chamber, instead of conventional water quenching, and recovers waste heat for power generation use. Shougang Group Corp. switched to the CDQ system during the period from fiscal year 1997 to fiscal year 2000 from the traditional coke wet quenching (CWQ) system. Therefore, energy consumption at its works was reduced and led to an energy-saving effect equivalent to 24,700 tonnes of crude oil per year. Convinced of the effectiveness of the CDQ technology from this result, the Chinese government referred to the introduction of CDQ equipment in its Comments on the Implementation of Ten Key Projects in the 11th Five-Year Program for National Economic and Social Development starting in 2006.

In September 2009, 140 delegates from the Japan-China Economic Association visited China and met with Chinese Premier Wen Jiabao. He expressed his belief that, on the basis of the bilateral relations already established in the past, relations between the two countries could be further promoted. He presented his basic view that the green economy should be further developed to serve as the key to economic exchange between the two countries and positioned energy savings and the environment as the most pressing issues.

4.8 CONCLUSION

In January 2007, the Cebu Declaration on East Asian Energy Security was signed by 16 nations — the 10 member countries of the Association of Southeast Asian Nations (ASEAN), Australia, China, India, Japan, Korea, and New Zealand. The countries have agreed to promote energy security and listed a series of goals aimed at providing "reliable, adequate and affordable energy supplies" for strong and sustainable economic growth and competitiveness. The goals include improving the efficiency and environmental performance of fossil fuel use, reducing dependence on conventional fuels through intensified energy efficiency and conservation programmes, and mitigating greenhouse gas emissions through effective policies and measures, thus contributing to global climate change abatement.

On 24 September 2009, Prime Minister Yukio Hatoyama delivered an address at the United Nations General Assembly session and proposed to build an East Asian Community to consolidate cooperation among Asian nations. In his speech, he said that "today, there is no way that Japan can develop without deeply involving itself in the Asia and the Pacific region". In close coordination and cooperation with the Asian and Pacific countries, Japan can also help itself by reducing the effects of climate change and by exporting Japanese technology and know-how. He went on to say that

> [he] looks forward to an East Asian community taking shape as an extension of the accumulated cooperation built up step by step among partners who have the capacity to work together, starting with fields in which we can cooperate — Free Trade Agreements, finance, currency, energy, environment, disaster relief and more.

Since the Hatoyama administration has been in office, relations between Japan and China have rapidly improved. Prime Minister Hatoyama, who pledged not to visit the Yasukuni Shrine, said he stands by then Prime Minister Tomiichi Murayama's official 1995 apology for Japan's colonialism and military aggression. At the summit meeting with Chinese President Hu Jintao in September 2009, Prime Minister Hatoyama stated that he hoped to "build the community based on the relationship of trust which will be realised by both Japan and China acknowledging the differences between the two countries". Chinese President Hu made five proposals: increasing the frequency of visits of leaders of the two countries, strengthening and developing economic and trade relationships, improving public sentiments, cooperating in dealing with international issues including those concerning Asia, and resolving the differences between the two countries in an appropriate way. It seemed possible that, with the Hatoyama administration having taken office, both Japan and China would strive to advance bilateral ties with a focus on energy saving and environmental cooperation.

At the meeting with Chinese President Hu Jintao, Prime Minister Hatoyama asked the Chinese side for their cooperation toward the establishment of an East Asian Community. However, it should be noted that, while the size of the Japanese and Chinese economies is more or less the same, there is a big gap between the two countries in terms of GDP per capita. There exist differences between the two countries with regard to the political structures, as China adopts the socialist system under the leadership of the Communist Party of China. Furthermore, on the security side, potential tensions still remain between China and the Japan-US alliance. Hatoyama indicated that the US is a potential member of his envisaged regional grouping, and it will, therefore, be essential to liaise with the United States and others to implement the concept of the community.

CHAPTER 5

INDIA'S PIPELINES: PARADOX, PROBLEMS AND POSSIBILITIES

Marie Lall

India's principal foreign policy aim has always been to be a recognised global power. From independence onwards foreign policy was developed in line with Nehruvian ideals and rooted in the paradigm of a multi-polar world. Over time and especially since the economic reforms after 1991, the methods to achieve this aim have changed — but the objective remains the same. Today economic growth and a close relationship with the US have been driving foreign policy — and both in turn are linked to energy security. This chapter will discuss how India's search for energy has shaped and changed Indian foreign policy making, both at a regional and at a global level, by focusing on three proposed transnational gas pipelines. It will also discuss how the search for energy, in particular for natural gas, has been hampered by domestic and international factors.

5.1 THE CENTRALITY OF THE QUESTION OF ENERGY SECURITY

India needs energy security first and foremost to ensure economic growth at 8–9% and at a rate which outstrips population growth, even in the current economic slump. India's growing population is expected to reach 1,180 million by 2010, 1,362 million by 2020 and 1,573 million by 2030, i.e. a more than 50% increase in less than 30 years. Since India is an energy supply–constrained economy, the sustainability of development and growth will depend on the availability of affordable, adequate and reliable energy and massive investments in social and physical infrastructure. Compared to its neighbours India's projected energy needs are huge (Figure 1).

Indian primary energy demand will grow proportionately to both population growth and economic growth. Coal is the largest source, constituting 51% of the total primary energy basket; the rest includes oil (36%), gas (9%) and nuclear and renewable (4%).[1] India currently imports 75% of its oil per annum, raising India's vulnerability to volatile oil markets. The Reserve Bank of India reports that every US$1 rise in the international price per barrel of crude oil adds US$600 million (around Rs 28 billion) to the country's import bill.[2] According to the International Energy Agency, a US$10 rise in crude prices would reduce India's GDP by 1%.[3] Hence there is an acute need to diversify the type as well as regional source of energy to ensure continuous

[1] M. Lall and I. Lodhi, "The Political Economy of the Iran-Pakistan-India Pipeline", ISAS working paper 26 (Institute of South Asian Studies, 2007), http://www.isn.ethz.ch/isn/Digital-Library/Publications/Detail/?ots591=0C54E3B3-1E9C-BE1E-2C24-A6A8C7060233&lng=en&id=96164.

[2] Paranjoy Guha Thakurta, "High Oil Prices Would Hit Indian Economy", *Business Line*, 20 March 2004, http://www.thehindubusinessline.com/2004/03/20/stories/2004032000240800.htm [Accessed 5 July 2007]; Asian Development Outlook 2006, *Economic Trends and Prospects in Developing Asia* (Manila: ADB, 2006).

[3] Federation of Indian Chambers of Commerce and Industry (FICCI), "Emerging Oil Price Scenario and Indian Industry", December 2004, http://www.ficci.com/surveys/FICCI-oil-survey-dec2004.pdf [Accessed 5 July 2007].

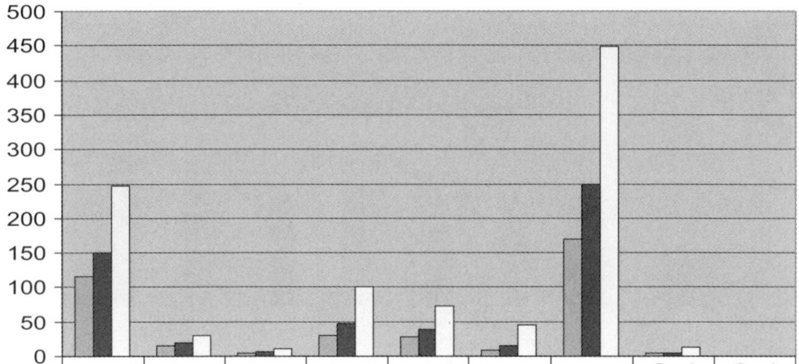

	Oil IN	Oil PK	Oil BDG	as IN	Gas PK	Gas BD	Coal IN	Coal PK	Coal BD
2003-04	116	15.2	3.71	29.74	27.29	8.29	169.9	3.3	0
2010	150.2	19.72	5.7	47.19	39.21	15.51	248.7	4.71	0.5
2020	246.9	30.94	11.6	101.88	72.75	44.03	447.6	13.9	0.9

Figure 1 Energy Needs (Oil, Gas and Coal) in India, Pakistan and Bangladesh
Source: M. Lall (ed.), *The Geopolitics of Energy in South Asia* (Singapore: ISEAS, 2009).

availability of energy at affordable prices.[4] Natural gas and nuclear energy have been the logical areas to expand and develop. With regard to increasing nuclear energy, in particular for electricity production, India negotiated the Indo-US civilian nuclear deal between 2007 and 2009, which now allows India to buy nuclear fuel, technology and other related materials despite India not being a signatory to the Non-Proliferation Treaty (NPT). India aims to expand the number of its reactors dramatically over the next 20 years. However, even a dramatic expansion is unlikely to yield enough energy to ensure the required economic growth.

---

[4]A slightly different scenario is painted by the US Energy Information Agency in whose reference case scenario the primary energy demand in India is expected to grow by 3.6% per year, doubling from 537 Mtoe in 2005 to 1,299 Mtoe in 2030. India will need to quintuple its electricity generation capacity from 1,600 GW to nearly 8,000 GW (A. Cohen, L. Curtis and O. Graham, "The Proposed Iran-Pakistan-India Gas Pipeline: An Unacceptable Risk to Regional Security", The Heritage Foundation, http://www.heritage.org/Research/AsiaandthePacific/bg2139.cfm [Accessed 8 February 2010]).

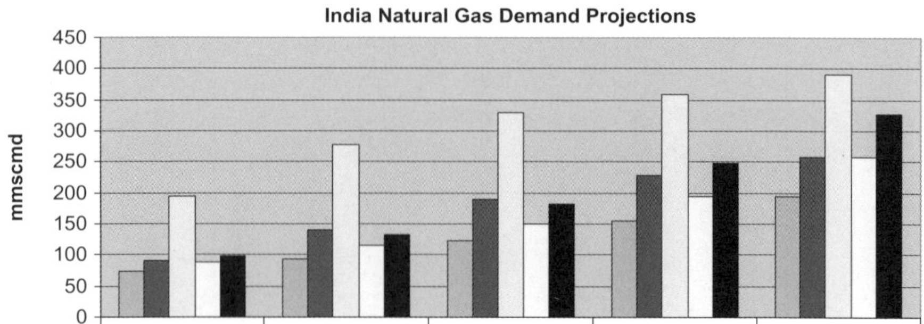

Year	2005	2010	2015	2020	2025
EIA	74	93	124	155	195
IEA	91	140	189	228	259
IHV	195	277	329	358	391
IV	89	115	149	194	258
PED	98	134	183	249	328

EIA IEA IHV IV PED

Figure 2 India Gas Demand Projections

Source: M. Lall and I. Lodhi, "Political Economy of Iran-Pakistan-India (IPI) Gas Pipeline", ISAS Working Paper No. 26, Institute of South Asian Studies, 23 October 2007, p. 9.

With regard to gas, India has had to look abroad to supplement local reserves, which despite recent discoveries are not sufficient. India's proven natural gas reserves are 38 Tcf (or 1.075 Tcm).[5] India produces 85 million metric standard cubic metres per day (MMscmd) or 1.08 Tcf per annum of natural gas.[6] The natural gas demand is expected to reach about 400 MMscmd by 2025 (Figure 2).[7]

---

[5] 1 trillion cubic feet (Tcf) = 0.0283 trillion cubic metres (Tcm).

[6] S. Srivastava, "India Grapples with Energy", *Asia Times*, 24 March 2007, http://www.atimes.com/atimes/South_Asia/IC24Df01.html [Accessed 5 May 2007].

[7] There have been two contentious and important issues in India in the context of gas and pipeline imports: demand/supply projections and pricing. Much of the controversy over demand/supply estimates, besides political motivations, arises from employing different econometric models and the inherent price sensitivity of the gas market. However, there is no substantial difference in the projections from different agencies except India Hydrocarbon Vision (IHV 2025). The IHV projections (see Figure 2) are substantially higher vis-à-vis other models, because they take into account the existing supply and demand gap which other models do not incorporate. The Eleventh (XI) Five-Year Plan

Given the need for increased natural gas, importing it became crucial. In February 2005 the Indian government approved plans for talks with six countries on the construction of gas pipelines that would pass through Iran, Afghanistan, Pakistan, Myanmar, Bangladesh and Turkmenistan. This new "pipeline diplomacy" was seen as a key foreign policy priority. However, the policy failed to bear the expected results. The next section will give details of the three pipelines which had been planned and the stage at which they are currently stuck.

5.2 THE PROPOSED PIPELINES

5.2.1 Iran-Pakistan-India (IPI) Pipeline

Iran has proven reserves of about 971.2 Tcf, the second-largest in the world after Russia. The IPI pipeline's source, the Pars field, contains 300 Tcf with a current production capacity of 3.1 Bcf/d. The Iran-Pakistan-India (IPI) pipeline was first discussed in the late 1980s. After four years of different studies, India signed an MoU in 1993 with Iran.[8] However, due to security concerns, especially with regard to the route through Pakistan, the project was shelved and only re-emerged later.[9] A agreement between heads of state was signed in June 2005 for the $4.5 billion pipeline project and in August 2005 the Indo-Iran joint working group met in New Delhi. The meeting in Islamabad in May 2006 saw the trilateral meeting at second secretary level.[10]

(2007–2012) acknowledges this gap which is currently managed by arbitrary rationing, resulting in an underutilisation of the installed capacity in the fertiliser as well as the power sector (Planning Commission, Government of India, Draft Report of Expert Committee on Integrated Energy Policy 2005, pp. 34, 49).

[8] S. Pandian, "The Political Economy of Trans-Pakistan Gas Pipeline Project: Assessing the Political and Economic Risks for India", *Energy Policy*, 33 (2005), pp. 659–670.

[9] India also explored the feasibility of offshore, deep-sea and shallow-sea pipelines as alternatives to crossing through Pakistani territory. However, the technological problems of a deep-sea (2,400 m) pipeline were difficult to overcome with the given technology. The shallow-sea option along Pakistan's coastal line, which would have had to cross Pakistan's exclusive economic zone (EEZ), was rejected by Nawaz Sharif's government due to Pakistani security concerns.

[10] "Gas pipeline from Myanmar likely to bypass Bangladesh", *Financial Express*, New Delhi, 16 May 2006, http://www.financialexpress.com/news/gas-pipeline-from-myanmar-likely-to-bypass-bangladesh/169582/1#.

The meeting discussed the technical, financial and legal aspects of the project besides issues of project structure, and a feasibility study including the route.[11] During his visit to Iran in May 2007, the Indian foreign minister Mr. Mukherjee reiterated the Indian commitment to the project.

With a total length of 2,775 km and at an estimated cost of $7.4 billion, the pipeline was meant to change the face of regional politics in South Asia. The gas pipeline would be expected to pump 60 MMscmd into Pakistan whilst India would receive 90 MMscmd. The pipeline was supposed to start from Assaluyeh, South Pars, stretching over 1,100 km, in Iranian territory before entering Pakistan and travelling through Khuzdar, with one section of it going south to Karachi on the Arabian Sea coast, and the main section travelling to Multan. From Multan the pipeline would link to Delhi. The pipeline would be running 1,115 km within Iran (Assaluyeh to the Pakistan border) and 898 km within Pakistani territory and another 740 km within India. A segmented construction approach had been agreed so that each country would build the pipeline within its own territory and would have proprietary rights. Concerns regarding gas price, transit fees and security were discussed bilaterally and trilaterally, and played their part in delaying and hampering the project. Yet the real obstacle arose through India's rapprochement with the US. In late April 2008, India and Pakistan held ministerial-level talks on both the Iran-Pakistan-India and the Turkmenistan-Afghanistan-Pakistan-India (TAPI) pipeline projects. India's Minister for Petroleum and Natural Gas Murli Deora said that both pipelines were equally important to Indian energy interests.[12] The TAPI project is discussed further, but for all intents and purposes the IPI project seems to have been shelved. In the meantime Iran and Pakistan have continued working on a pipeline without Indian cooperation.

5.2.2 Turkmenistan-Afghanistan-Pakistan-India (TAPI) Pipeline

Turkmenistan's gas reserves have been a debated subject for a number of years. For years Russia peddled the myth that there were insufficient

[11] "Accord on early IPI has pipeline work", *Dawn the Internet Edition*, Islamabad, 23 May 2006, http://archives.dawn.com/2006/05/23/top9.htm.

[12] Cohen, Curtis and Graham, 2010.

reserves to supply both Eastern and Western demand. However, in recent years, it has emerged that the reserves are indeed substantial, resulting in China pursuing a gas deal. (It is estimated that Turkmenistan has 10–14 Tcm of reserves, which are however not all certified.)

Also called the Trans-Afghanistan Pipeline, it was planned to transport Caspian Sea natural gas from Turkmenistan through Afghanistan into Pakistan and then to India. The Afghan government is expected to receive 8% of the project's revenue. The original project started in March 1995 when an inaugural MoU between the governments of Turkmenistan and Pakistan for a pipeline project was signed. By October 1997 Unocal established the Central Asian Gas Pipeline consortium to build the Turkmenistan-Pakistan segment of the pipeline at an estimated cost of US$2 billion (US$2.7 billion if extended to India). Construction was scheduled to begin as early as 1998, but the ongoing civil war in Afghanistan obstructed any opportunities for financing the project. After the withdrawal of two major financers, Russian Gazprom and Unocal, the consortium was dismantled. After the United States invasion of Afghanistan in 2001, the TAPI project received a new impetus with the removal of the Taliban regime in Afghanistan, following a summit between the leaders of Turkmenistan, Afghanistan and Pakistan in May 2002. The parties affirmed their decision to revive the project and established a Steering Committee to oversee its implementation. At its first meeting in July 2002, the Steering Committee invited the Asian Development Bank (ADB) to play the role of development partner and help prepare the feasibility study for the project. In December 2002, the ADB approved a $1 million technical assistance grant for undertaking a feasibility study. India was officially invited in 2005, and India agreed.[13]

The 1,680 km (1,040 mi) pipeline would run from the Dauletabad gas field to Afghanistan. From there the TAPI pipeline would be constructed alongside the highway running from Herat to Kandahar, and then via Quetta and Multan in Pakistan. The final destination of the pipeline would be the Indian town of Fazilka, near the border between Pakistan and India. The cost of the pipeline was estimated at US$7.6 billion. Eleven high-level planning meetings have been held during the past

[13] M. Lall (ed.), *The Geopolitics of Energy in South Asia* (Singapore: ISEAS, 2009).

seven years, with Asian Development Bank sponsorship and multilateral support.[14]

In April 2008, Pakistan, India and Afghanistan signed a framework agreement to buy natural gas from Turkmenistan. India discussed the pipeline project in September 2008 during meetings External Affairs Minister S. M. Krishna had with Turkmenistan President Gurbanguly Berdimuhamedow, Deputy Prime Minister in charge of oil and gas Baymyrat Hojamuhammedow and Foreign Minister Rasit Meredow in Istanbul.[15] Given that the TAPI project is supported by the US in light of the Afghanistan reconstruction, there is more political will behind the project than with the IPI project. However, there are other issues and misgivings about the TAPI project which are discussed in the next section.

5.2.3 Myanmar-Bangladesh-India Pipeline

Myanmar has large gas reserves — according to data released by the Myanmar Oil and Gas Enterprise (MOGE) recoverable natural gas resources are around 51 Tcf due to two large offshore fields opposite Thailand and Bangladesh.[16] The first offshore gas project, Yadana, was developed at a cost of $1.2 billion off Moattama (blocks M5 and M6). It started to export gas to Thailand in 1988. The second, Yetagun, has cost about $700 million and is off the Tanintharyi coast in the southeast (Blocks M12, M13 and M14). The reserves are estimated at 3.2 Tcf. The third site is called Shwe, off the Rakhine coast, and is just being developed. It could yield around 4.2–5.8 Tcf.

Today India's interests lie in A-1 and A-3, offshore blocks in Shwe. India's state-owned ONGC Videsh Limited (OVL) holds 20% and the

---

[14] A Review, "Gas-rich Turkmenistan and TAPI Pipeline May Bring Peace to Kabul", http://areview.co.cc/mfarhanonline/featured/gas-rich-turkmenistan-and-tapi-pipeline-may-bring-peace-to-kabul/ [Accessed 5 February 2010].

[15] "Krishna Discusses TAPI Pipeline Project with Turkmenistan", *The Economic Times*, http://economictimes.indiatimes.com/news/news-by-industry/energy/oil-gas/Krishna-discusses-TAPI-gas-pipeline-project-with-Turkmenistan/articleshow/5031220.cms [Accessed 5 February 2010].

[16] Tin Maung Maung Tan, "Myanmar's Energy Sector: Banking on Natural Gas", in *Southeast Asian Affairs 2005*, Singapore: Institute of Southeast Asian Studies, 2005, pp. 257–289.

Gas Authority of India (GAIL) and KoGas (Korea Gas Corporation) each hold 10% of the equity stakes in both the A-1 and A-3 blocks, which are close to India. Daewoo International holds 60% of the stakes in the project. The exploration rights for blocks A-2 and L have been acquired by Essar, a private firm. Further deep-sea blocks have been auctioned to China and India and await certification and development.

As of 2005 India was planning to build a pipeline to bring Myanmar gas to eastern India. Several routes were considered — the most likely ones through Bangladesh, although at first the underwater option was considered and then dropped due to the high cost levels. Both routes through Bangladesh, one following the Kaladan River and travelling through Tripura and Mizoram, and one entering Bangladesh without going through India, were costed at $1 billion.

A pipeline between Myanmar and India always depended on the quantity of gas available on the Rakhine coast, as well as the cost of bringing the gas back to India. The end-user price was also an issue as the price would reflect the cost of laying the pipeline and other costs such as transit fees. Whilst price and security were genuine issues, the dialogue went astonishingly well. India had arrived at the penultimate stage and was getting the framework signed. Bangladesh's extra demands remained problematic but the transit fee would have mitigated the balance of payments issue. The issue of the trade and energy corridor between Bangladesh, Nepal and Bhutan delayed India's final offer to Myanmar. Mani Shankar Aiyar's plans included a special diplomatic effort so that the chief ministers of the northeast could improve relations with Bangladesh and restore the status quo ante of 1965 with East Pakistan when goods could be traded freely. Due to the negotiation problems with Bangladesh, another route through India's northeast, bypassing Bangladesh, was also studied. The cost would have been $3 billion.[17]

However, as New Delhi took time to get back to MOGE, China got involved. China's PetroChina signed a gas export memorandum of understanding with Myanmar early in 2006 and completed the survey for a

---

[17] M. Lall, "Indo-Myanmar Relations, a Shifting Geopolitical Scenario", ISAS working paper 29 (Institute of South Asian Studies, 2007), http://www.isasnus.org/events/working-papers/28.pdf; and Lall (ed.), 2009.

2,389 km pipeline from Kyaukphyu (which will be developed as a deep-sea port) in Myanmar to China's Yunnan Province. The MoU envisages 6.5 Tcf being supplied from A-1 for 30 years. In return China will make available an $84 million soft loan.[18]

There is not enough gas for both China and India from both these blocks. The in-place reserves from the A-1 block in the Shwe field have been assessed by Houston-based consulting firm Ryder Scott at 2.88–3.56 Tcf.[19] Today the deal lies shelved although the press is still talking about a potential pipeline through Mizoram linked to the Kaladan multimodal transport project. It will in essence depend on the amount of gas available from the Essar Oil's stakes in Myanmar's offshore block A-2 and onshore block L. Essar hopes to start production by 2013.

5.3 PROBLEMS WHICH HAVE FRUSTRATED THE DEVELOPMENT OF THE PIPELINES

Despite India's need for natural gas and negotiations stretching over years, the three pipelines have not come to fruition. Reasons for this are to be found both in the foreign policy as well as the domestic realm. This section will explore how India's foreign relations have contributed to frustrating the projects before analysing the domestic issues which have also played a role. It is also clear that India's foreign relations have been profoundly marked by the search for energy security — in most cases to the detriment of good regional and neighbourly relations. The balance of power is shifting towards China as it sucks up all regional energy resources which can be piped back over land. The shift from self-sufficiency to competitive alliance politics is leading to a new great game encompassing the whole of South Asia with the US and China, but increasingly Russia is playing out its global power politics in the region as well.

---

[18] Sanjay Dutta, "Gas pipeline: Myanmar takes India for a ride", *Times of India*, 26 March 2006, http://articles.timesofindia.indiatimes.com/2006-03-26/india-business/27796160_1_gas-pipeline-china-s-yunan-export-pipeline.

[19] Anand Kumar, "India-Myanmar Gas Pipeline: Disentangled at Last", Paper no. 1822 (South Asia Analysis Group, 2006).

5.3.1 Foreign Policy Issues

More than anything else the Manmohan Singh Congress–led government staked its foreign policy around the improvement of relations with the United States. Central to this was the negotiation of a civilian nuclear deal allowing India to buy nuclear fuel and technology despite not being a signatory to the NPT. The deal was finally agreed in 2008 with some follow-up legislation such as the Nuclear Liability Act still being put in place in 2010. The conditions associated with the deal have had an impact on the "independence" of India's foreign policy decision making as the 123 Agreement specifically states that the deal can be cancelled if either state undertakes any action seen by the other as affecting the other's national security. This has given the US a stake in Indian decision making and influencing India on its Iran policy has been a priority. The US has put pressure on India not to buy natural gas from Iran, which it considers a pariah state (or a member of the "axis of evil").[20] The role of the US has had further impacts on the development of the IPI pipeline, in particular with regard to the financing. This has proved problematic due to the US sanctions regime on Iran. Yet despite the Iran and Libya Sanctions Act of 1996 and the pending Iran Counter-Proliferation Act of 2007 by the US Congress, companies have been doing business with Iran much beyond the limit of $20 million mentioned in these Acts. A consortium was also under discussion among BHP (Australia), NIGC (Malaysia), Total (France), Shell (Netherlands) and BP (UK) in addition to Iranian, Pakistani and Indian national gas companies.[21] Russian Gazprom has also expressed interest despite the sanctions regime, especially since Russia would prefer the gas not to flow to Europe. Despite all this economic goodwill the financing has not come together, and having successfully influenced

---

[20]In January 2006, the US ambassador to India explicitly linked progress on proposed United States–India civil nuclear cooperation with India's upcoming vote against Iran, saying if India chose not to side with the United States, he believed the US-India initiative would fail in Congress. The United States officials repeatedly and unequivocally expressed their opposition to the IPI gas pipeline.

[21]S. Shahid, "Iran-Pak-India Gas Pipeline: Implications and Prospects", *Business & Finance Review*, 15 January 2007.

New Delhi to put the deal on the backburner, the US is now leaning on Pakistan not to buy gas from Iran either. The TAPI pipeline on the other hand is part of Afghanistan's 2008 budget and has had US support since the 1990s. The aim is to stabilise Afghanistan and is therefore being pushed as a viable alternative for India's gas supply by US policy makers.

Relations with India's neighbours were supposed to improve dramatically through the interdependence created through the pipelines — this was particularly expected with the IPI pipeline. Tying India's energy security in with Pakistan was expected to help with the peace process through increased trade as well as help boost the confidence-building measures. However, issues which were not resolved between Pakistan and India included the transit and tariff fees levied by Pakistan for the IPI pipeline. Pakistan initially asked $1 per MBtu as a transit fee and $1.57 per MBtu for transportation (and tariff) while India was not willing to pay more than $0.15 and $0.40 respectively. Pakistan offered India the alternative option to buy gas at the Pakistan-India border from Pakistan and let Pakistan and Iran deal with the pipeline, but India felt this would give Pakistan too much leverage.[22] Beyond leverage, security in Pakistan was of paramount importance to India. The IPI pipeline would pass through Balochistan where an insurgency has already led to attacks on the Pakistani pipeline grid. New Delhi's security establishment is not confident that Pakistan can guarantee the security of a pipeline in that region. There are similar issues between both countries with the TAPI pipeline but they are yet to be addressed.

Indo-Bangladesh relations have also been strained in the past and energy interdependence was seen as a way to balance out India's hegemonic position. The proposed pipeline from the Shwe gas field to Kolkata would have earned Bangladesh about $125 million annually in transit fees, improving the balance of payments difference between the two countries. Dhaka had agreed in principle and a tripartite deal was to be sealed in the course of 2005. The Bangladeshi government wanted the pipeline to be laid along Bangladesh's existing roads and highways to make it

[22] S. Srivastava, "India Weighs the Pipeline Odds", *Asia Times*, 19 April 2007, http://www.atimes.com/atimes/South_Asia/ID19Df01.html [Accessed 5 May 2007].

easier to manage. It also wanted India to allow it to use the pipeline to export gas to India or import gas from Myanmar. Whilst India would be responsible for building the $1 billion gas pipeline, Bangladesh's state-owned Gas Transmission Company would be responsible for managing the part of the pipeline based in its country. In July 2005 the Bangladeshi government made three further demands: the provision of transit facilities through India to facilitate transmission of hydro-electricity from Nepal and Bhutan to Bangladesh; to be allowed to utilise the corridor for trading between Bangladesh and Nepal/Bhutan through Indian territory; and an initiation of measures to reduce the $2 billion trade imbalance between Bangladesh and India. The Indian government opposed these conditions, as they did not want to make bilateral issues part of a trilateral agreement. Given this, the possibility of bypassing Bangladesh by taking the pipeline from Myanmar into Mizoram and onwards to Assam and culminating in West Bengal, a distance of 1,400 km, was explored as an alternative. In the end Mani Shankar Aiyar, the minister responsible at the time, managed to negotiate a deal with Bangladesh; however, both the Indian Army and the Ministry of External Affairs vetoed the pipeline through Bangladesh on security grounds. The whole process had taken so long that in the meantime China had made an alternative offer to Nay Pyi Taw and both countries agreed that the gas from A-1 and A-3 of the Shwe fields would be piped to Kunming instead. This was not the first time that India had lost a gas deal to China and it has not helped improve Indo-Chinese relations.

Relations between India and China have been problematic since the 1962 war, which India lost. Since then China has been seen as the only "real" threat that India faces. There has been fierce competition for energy resources, especially in Central Asia and Myanmar.[23] Indian and Chinese firms have cooperated in some cases, in order to avoid driving up the price in the bidding war; however, this has only ever been the case when the gas was "overseas" and could not be repatriated to China over land. China's ability to pay higher prices and for its representative teams to close deals with governments without having to go back to Beijing also give

[23] M. Lall, "The Geopolitics of Energy in Asia: Indo-Chinese Competition over Myanmar's Gas Reserves", *Panorama*, 2 (2007), pp. 35–47.

the country an edge over its Indian counterparts. China's hydrocarbon companies also have political muscle in Beijing, unlike their Indian counterparts in Delhi.[24] China not only secured Myanmar gas but has also expressed an interest in Iranian gas from the Pars field (originally intended for the IPI pipeline) and signed a deal with Turkmenistan for gas which is earmarked for the TAPI pipeline. India now feels that China's string of pearls strategy means it is surrounded by Chinese allies. Relations have worsened with the arguments around the border of Arunachal Pradesh and China's disapproval of the Indo-US nuclear deal.

Relations with India's other neighbour, Myanmar, remain good, despite the fact that the pipeline was not realised. India's aim still is to cooperate with the Myanmar government in anti-insurgency campaigns on the northeastern joint border as well as to open up a trade corridor with Southeast Asia, tying itself closer to the ASEAN countries. India's energy companies are still involved in various gas fields — both the original ones and some new ones — and there is still hope in Delhi that with new gas finds a pipeline could be viable one day. This in turn would strengthen Indo-Myanmar ties.

Further afield relations for India have also become problematic — whilst Iran and India have had improved relations since the Islamic revolution, India's increasingly pro-American stance has complicated things. In the early 1990s Iran and India both supported the Northern Alliance as opposed to Pakistan's support for the Taliban in Afghanistan. Later both countries supported the United States military–led ousting of the Taliban regime. India and Iran have more recently cooperated in the Afghanistan reconstruction activities. Tehran sees India and China as its potential customers for oil and gas as well as a way out from the US attempts to isolate Iran. The years 2003–2005 saw a further deepening of India-Iran ties. The IPI pipeline discussions were renewed in the same period and a US$40 billion LNG deal was signed to supply 7.5 million tonnes of liquefied natural gas (LNG).

The recent nuclear deal between the United States and India has put some pressure on India-Iran relations. However, prior to the nuclear deal,

[24] Bo Kong, "The Geopolitics of the Myanmar-China Oil and Gas Pipelines", *Pipeline Politics in Asia: The Intersection of Demand, Energy Markets and Supply Routes*, The National Bureau of Asian Research, NBR Special Report #23, (September 2010), pp.55–67.

there were two major issues regarding the IPI pipeline: the price of gas and the financing of the construction of the pipeline. Gas prices proposed by Iran were more than double what Pakistan and India were willing to accept. India wanted to pay a fixed amount per unit delivered to its border, but Iran wanted the cost to be linked to the fluctuating international energy prices, saying the prices offered by Pakistan and India were only half of what it was looking for.[25] Given that the IPI pipeline is at least provisionally shelved, relations are sensitive as India is increasingly seen as taking the US side on the international stage.

India has been, for a number of years now, interested in increasing its influence in Central Asia. Whilst relations with Iran are more complicated these days, India has improved its relationship with post-Taliban Afghanistan by becoming heavily involved in the reconstruction efforts. India is keen to maintain good relations as a counterbalance to Pakistan and fears a return of the Taliban once the US and coalition troops leave, whom they believe would be in alliance with Pakistan. A pipeline through Afghanistan is not only favoured by Washington but would also further strengthen Indo-Afghan relations. However, given the current security situation and the terrain the pipeline would have to traverse, such a pipeline seems to be more of a political plan rather than an impending reality. Beyond the situation in Afghanistan there are also issues pertaining more directly to Turkmenistan and the availability of Turkmen gas. There is a huge issue around the question of whether Turkmenistan has overcommitted its gas supplies:

> Russia has a virtual monopoly over exports from Turkmenistan with Gazprom allowed to take up to 50 bcm of gas annually for another two decades. China has been promised 30 bcm from 2009 and Iran 8 bcm. Berdymukhamedov is

---

[25] Price negotiations in the IPI context are quite complex. Iran had initially demanded $7.20 per million British thermal unit (MBtu) linked with a price escalating component (10% of Brent Crude Oil) that was almost double the $4.25 which India offered on its border. Both Pakistan and India rejected this offer, after which a consultancy firm, Gaffney, Cline & Associates, was hired to give a workable formula. The local gas production price in India and Pakistan is approximately $3 after subsidies. According to the new formula the price of gas was to translate to $4.93 per MBtu that is linked to the Japan Crude Cocktail price.

Khaleeq Kiani, "ECC approves price formula: Work on IPI pipeline to begin next month", *Dawn the Internet Edition*, 11 April 2007, http://archives.dawn.com/2007/04/11/top1.htm.

also reported to have signed another 10 bcm to the European Union, bringing its projected gas exports to over 100 bcm annually.[26]

Russia is heavily involved in the Turkmenistan gas deals and would prefer to see India opt for the IPI as opposed to the TAPI pipeline. Russia remains the most influential player on the Turkmenistan stage and India has now to contend with both Chinese and Russian policy if it wants to source gas for the TAPI pipeline.

5.3.2 Domestic Constraints

Looking at the complex foreign policy issues, it is easy to forget that domestic policies also play a role in securing energy resources. India's prime problem has been the lack of a long-term energy security vision as well as the coordination of the various ministries and government departments to secure such a vision. Given the democratic structure and the devolved state government power base where local priorities play an increasingly important role, domestic opposition at the local, regional and national levels has also played a part.

India's lack of vision and failure to prioritise is very different from China's world plans. Since India is a democracy all things are subject to negotiation and compromise. This has been increasingly so since the mid-1990s when coalition governments have ruled in New Delhi. As a result policies rarely are sustained beyond the term of one administration and likely to fall prey to interparty negotiations even within a term. Whilst India has a clear vision of its position in the world it has not yet developed a strategy to succeed in this aim. Part of the problem is governmental and ministerial non-coordination and non-cooperation. This is exemplified by the failure of the India-Myanmar pipeline where the various ministries could not agree — that is if they were even talking to each other.[27] The Ministry of External Affairs and the Ministry of Defence were against the project, citing security reasons, whilst the Ministry of Commerce and the Department

[26] L. Jishnu, "Is IPI Pipeline Less Risky than TAPI Line?", Rediff, 31 May 2008, http://www.rediff.com/money/2008/may/31guest.htm [Accessed 6 February 2010].

[27] Lall, 2007.

for Northeastern Affairs were for the pipeline. The part of the Energy Ministry dealing with gas[28] was caught in the midst of a turf war. During this time the Prime Minister's Office was negotiating the nuclear deal and the pipeline from Myanmar was therefore not on the high priority agenda. Consequently China was able to step in and negotiate a deal.[29]

Beyond the ministries there is domestic opposition to politically sensitive engagement (with an active press keeping the level of public opinion and information raised) as well as ambivalence across domestic political circles.

5.3.3 The Future: The Need for a Policy Turnaround

Given the failure of securing three transnational pipelines and the lessons hopefully learnt in the process, India's energy policy is in dire need of revision. The Indian government needs to develop a strategy which matches a long-term vision (20+ years) linking economic and political drivers with regard to relations in the region and across the globe. Domestic ambivalence and the focus on winning the next elections will always mean that India will lose out to China. Energy security policy has to be at the heart of this. As a result different ministries need to work together with the same goals and visions. The Prime Minister's Office cannot be the only place where strategic decisions are made and pushed through; it also needs to lead the other ministries and harmonise their working with the government. Party politics and regional politics need to be put on the backburner when dealing with such complex issues which are however of prime national importance. Beyond domestic reform both the Ministry of External Affairs and the Indian Army need to gain a more in-depth understanding of China's modus operandi in order to develop a long-term plan for what to do regarding China and China's encroaching presence in the region.

---

[28]The fact that India's Ministry of Energy has been split into three ministries is not of help either.

[29]Lall, 2007.

BIBLIOGRAPHY

A Review. "Gas-rich Turkmenistan and TAPI Pipeline May Bring Peace to Kabul", http://areview.co.cc/mfarhanonline/featured/gas-rich-turkmenistan-and-tapi-pipeline-may-bring-peace-to-kabul/ [Accessed 5 February 2010].

Asian Development Outlook 2006. *Economic Trends and Prospects in Developing Asia*. Manila: Asian Development Bank, 2006.

Cohen, A., Curtis, L. and Graham, O. "The Proposed Iran-Pakistan-India Gas Pipeline: An Unacceptable Risk to Regional Security", The Heritage Foundation, http://www.heritage.org/Research/AsiaandthePacific/bg2139.cfm [Accessed 8 February 2010].

Federation of Indian Chambers of Commerce and Industry (FICCI). "Emerging Oil Price Scenario and Indian Industry", December 2004, http://www.ficci.com/surveys/FICCI-oil-survey-dec2004.pdf [Accessed 5 July 2007].

Jishnu, L. "Is IPI Pipeline Less Risky than TAPI Line?", Rediff, 31 May 2008, http://www.rediff.com/money/2008/may/31guest.htm [Accessed 6 February 2010].

"Krishna Discusses TAPI Pipeline Project with Turkmenistan", *The Economic Times*, http://economictimes.indiatimes.com/news/news-by-industry/energy/oil-gas/Krishna-discusses-TAPI-gas-pipeline-project-with-Turkmenistan/articleshow/5031220.cms [Accessed 5 February 2010].

Kumar, A. "India-Myanmar Gas Pipeline: Disentangled at Last", Paper no. 1822, South Asia Analysis Group, 2006.

Lall, M. "Indo-Myanmar Relations in the Era of Pipeline Diplomacy", *Contemporary South East Asia*, vol. 28, no. 3 (2006), pp. 424–446.

Lall, M. and Lodhi, I. "The Political Economy of the Iran-Pakistan-India Pipeline", ISAS working paper 26, Institute of South Asian Studies, 2007, http://www.isn.ethz.ch/isn/Digital-Library/Publications/Detail/?ots591=0C54E3B3–1E9C-BE1E-2C24-A6A8C7060233&lng=en&id=96164.

Lall, M. "Indo-Myanmar Relations, a Shifting Geopolitical Scenario", ISAS working paper 29, Institute of South Asian Studies, 2007, http://www.isas-nus.org/events/workingpapers/28.pdf.

Lall, M. "The Geopolitics of Energy in Asia: Indo-Chinese Competition over Myanmar's Gas Reserves", *Panorama*, 2 (2007), pp. 35–47.

Lall, M. (ed.) *The Geopolitics of Energy in South Asia*. Singapore: ISEAS, 2009.

Pandian, S. "The Political Economy of Trans-Pakistan Gas Pipeline Project: Assessing the Political and Economic Risks for India", Energy Policy, 33 (2005), pp. 659–670.

Planning Commission, Government of India. Draft Report of Expert Committee on Integrated Energy Policy, 2005.

Shahid, S. "Iran-Pak-India Gas Pipeline: Implications and Prospects", *Business & Finance Review*, 15 January 2007.

Srivastava, S. "India Grapples with Energy", *Asia Times*, 24 March 2007, http://www.atimes.com/atimes/South_Asia/IC24Df01.html [Accessed 5 May 2007].

Srivastava, S. "India Weighs the Pipeline Odds", Asia Times, 19 April 2007, http://www.atimes.com/atimes/South_Asia/ID19Df01.html [Accessed 5 May 2007].

Thakurta, P. G. "High Oil Prices Would Hit Indian Economy", *Business Line*, 20 March 2004, http://www.thehindubusinessline.com/2004/03/20/stories/2004032000240800.htm [Accessed 5 July 2007].

Tin Maung Maung Tan, "Myanmar's Energy Sector: Banking on Natural Gas", in *Southeast Asian Affairs 2005*, Singapore: Institute of Southeast Asian Studies, 2005, pp. 257–289.

CHAPTER 6

THE CAUCASUS: CONFLICT, INSTABILITY AND FOSSIL ENERGY EXPORT ROUTE

Hooman Peimani

6.1 INTRODUCTION

As a bridge between Asia and Europe, the Caucasus[1] has historically been important for regional and non-regional powers. Especially over the past 3,000 years, this region has been a scene of rivalry between and among these powers seeking to dominate or control it in one form or another. In addition to its agrarian, mineral and energy (mainly oil) richness, its major attraction for them has been its importance as an intercontinental land link. This geographical reality is a major contributing factor to its strategic significance. The latter has become especially prominent in the post-Soviet era when the regional countries, namely Armenia, Azerbaijan and Georgia, gained independence after about two centuries of forced

[1] Throughout this chapter, the Caucasus refers to Azerbaijan, Armenia and Georgia. Hence the Russian North Caucasian republics (e.g. Chechnya, Dagestan and Ingushetia) are excluded.

membership in the Russian and Soviet states. That the Caucasus neighbours Iran and Russia, two sources of concern for the Western countries for different reasons, has also made this region crucial for their "containment" policy.

The post-Soviet era has not been peaceful for any of the Caucasus's forming states. On the contrary, to a varying extent, intra- and inter-state wars have hurt them all, adding new problems to those political, economic, social, military/security and environmental ones inherited from the Soviet era. Georgia experienced two devastating civil wars instigated by the bid for independence of its two breakaway republics of South Ossetia and Abkhazia. Respectively, they ended in 1992 and 1993, only to leave those republics practically independent, a status which has lasted to this day (February 2010). Begun in the last years of the Soviet era, the efforts of the Armenian Karabakhis to separate their region (Nagorno-Karabakh) from Azerbaijan and reunite this predominantly Armenian region with Armenia instigated a civil war in Azerbaijan upon its independence in 1991; it pitted Azerbaijan against Armenia, the main backer of the Armenian Karabakhis. A ceasefire agreement ended the conflict in 1994 without addressing its main cause. The Armenian Karabakhis have since controlled that region and the adjacent land connecting it to Armenia. About 20% of the Azeri territory has since been out of Baku's control.

The Caucasus has also experienced a major inter-state war in its independence era. The August 2008 military attack of the Georgian government on South Ossetia did not bring about the desired result for Tbilisi. Instead of securing its control over the breakaway republic, it provoked a Russian military response in support of its protégé. The Russian military devastated the Georgian military and destroyed almost the entire Georgian military industry throughout the country. The development provided the opportunity for Moscow to recognise Abkhazia's and South Ossetia's declared independence as demanded by their governments.

The Caucasus has therefore had more than its fair share of armed conflicts and instability since its independence. Today the region is not engulfed in any armed conflict or major internal unrest affecting its forming states, but it is also far from becoming a stable region. In fact, it is now prone to instability as Azerbaijan's and Georgia's territorial loss and the continuity of a no-war-no-peace situation between them and their breakaway regions have planted the seeds of armed conflicts in the region. To this also should be added the prevailing hostile relations between Georgia and Russia over the latter's backing of the former's breakaway republics and of course the humiliating memory of a military defeat in Georgia. To put it briefly, the situation is ripe for a new round of intra- and inter-state armed conflicts. Their outbreak could well drag in the regional (Iran and Turkey) and non-regional (the US and the European Union (EU)) powers in support of one side or another, given their established pattern of relations with the Caucasian countries.

Although independence has not brought peace and prosperity for the Caucasus, this region could potentially have a much better future. In part, this is because of Azerbaijan's fossil energy resources, particularly oil, whose exports have provided since the second half of the 1990s a degree of financial strength for Azerbaijan, the only regional country with significant oil and gas reserves. However, as Azerbaijan's fossil energy deposits are not large enough to grant it a long-term export capability at a significant scale, the energy-related prosperity of the region is linked to its functioning as a major transit route for the Caspian region's oil and gas exports. It should therefore play this role for Kazakhstan and Turkmenistan, two Caspian countries with, respectively, substantial oil and gas reserves that ensure their export capability for two or three decades depending on the volume of their annual exports.

Thanks to its geographical location, this possible role if it becomes a reality in a sustainable manner creates an economic ground for a more prosperous future for the Caucasus. In fact, the efforts to that effect, translated particularly into oil/gas pipeline construction since the 1990s, have made this possibility more realistic. However, among other factors, there are doubts about this scenario coming true owing to the aforementioned sources of instability and armed conflicts in addition to the internal sources of social and political instability, particularly in Azerbaijan and Georgia. Consequently, the long-term oil/gas-related prosperity of this region will require peace and security, which, in turn, demand settling in a satisfactory manner for all concerned parties the existing inter- and intra-state conflicts. Against this background, certain factors capable of preventing the region's desired export role warrant elaboration.

6.2 OIL AND THE CAUCASUS

The importance of oil for the Caucasus is twofold. On the one hand, the region is rich in oil thanks to Azerbaijan's substantial oil reserves, the bulk of which is concentrated in its Caspian offshore oil fields. This reality has been the single major reason for the interest of the Western countries in Azerbaijan, especially the United States. Along with many European oil companies, American oil companies have practically dominated the fossil-energy sector of Azerbaijan, like those of other newly emerged Caspian oil exporters: Kazakhstan and Turkmenistan. Azerbaijan's proven oil reserves are estimated to be 7 billion barrels.[2] On the other hand, the Caucasus could play a major role as a transit route for oil and also gas exports from the Caspian region.

The entire Caspian region has a much smaller amount of recoverable oil than suggested by the very unrealistic estimates of the 1990s, e.g. over 200 billion barrels.[3] Its proven oil reserves, including those of Azerbaijan, Turkmenistan and Kazakhstan, excluding the non-Caspian oil reservoirs of the two other Caspian littoral states (Iran and Russia), is a fraction of those estimates, i.e. between 17 and 44 billion barrels, based on a 2006 estimate.[4] The recent discoveries have slightly pushed up its reserves to 47.4 billion barrels as of 2009.[5]

Within this context, Azerbaijan cannot possibly turn itself into a major long-term energy exporter given its fossil energy reserves (being mainly oil and, to a much smaller extent, gas) are too small. The evidence today substantiates this claim. Through various pipelines and by rail, its oil exports of about 211,000 barrels per day (bpd) in 2004[6] increased to

---

[2] BP, "Oil: Proved Reserves", *BP Statistical Review of World Energy* (London: BP, June 2009), p. 6.

[3] Bijan Mossavar-Rahmani, "The Challenge of the US Caspian Sea Oil Policy", *Motaellat-e Asyaie Markazi va Ghafghaz* [Central Asia and the Caucasus Review], 28 (Winter 2000), p. 51. Quoted in Hooman Peimani, *The Caspian Pipeline Dilemma: Political Games and Economic Losses* (West Port, CT: Praeger, 2001).

[4] Energy Information Administration, *Country Analysis Brief: Caspian Sea — Oil*, 2006, http://www.eia.doe.gov/emeu/cabs/Caspian/Oil.html [Accessed 10 December 2009].

[5] The figure is equal to the combined oil reserves of Azerbaijan, Kazakhstan and Turkmenistan as provided by BP, "Oil: Proved Reserves", p. 6.

[6] Energy Information Administration, *Country Analysis Brief: Azerbaijan*, June 2005, http://www.eia.doe.gov/emeu/cabs/azerbjan.html [Accessed 10 December 2009].

749,000 bpd in 2008, the most recent year on which statistics are available. At this small rate of export compared to the major international oil exporters such as Iran (about 2.6 million bpd), the country's oil reserves will be depleted in about two decades unless major new oil fields are discovered.

Currently, Azerbaijan has a much larger capacity for export because of its two major export pipelines (Baku-Supsa, in operation since the 1990s, and Baku-Tbilisi-Ceyhan (BTC), which went online in 2006). The BTC pipeline has provided additional capacity for the Azeris to export of about 1.2 million bpd, to be increased to 1.6 million bpd by the end of 2010 as announced in July 2009. As it is unlikely for Azerbaijan to achieve that export capability, it is necessary for the survival of the pipeline to be used also by other Caspian oil exporters, primarily Kazakhstan, which has the region's largest oil reserves of 39.8 billion barrels,[7] and Turkmenistan and Uzbekistan, which have small oil export capabilities. Yet, to date, there has not been any strong evidence that these exporters, especially Kazakhstan, are willing to commit themselves to export via Azerbaijan and particularly the BTC pipeline on a long-term basis and at a large scale. As the most recent example, in October 2009, Azerbaijan and Kazakhstan signed an energy agreement whose specifics are yet to be released. However, while such agreement could increase Kazakhstan's oil exports via Azerbaijan, at the end of 2009 only about 100,000 bpd of oil was moving "via the BTC pipeline and by rail to Georgia's Black Sea ports" through Azerbaijan.[8]

Against this background, Azerbaijan's capability to function as a new oil supplier for the global economy, whose energy requirements are increasing on a steady basis, currently qualifies it as an oil exporter of international significance, though a short-term one, despite its aforementioned characteristic. In particular, concerns over heavy reliance on the Persian Gulf fossil energy resources whose exports could be sharply reduced or cut due to a major conflict has made many economies, including the Western ones, China, Japan and South Korea, to seek to diversify

[7] BP, "Oil: Proved Reserves", p. 6.

[8] Energy Information Administration, *Country Analysis Briefs: Azerbaijan*, October 2009, http://www.eia.doe.gov/emeu/cabs/Azerbaijan/Full.html [Accessed 27 January 2010].

their energy suppliers. That objective has created interest in Azerbaijan as a supplier and also as a backup source for those depending on the Persian Gulf, mainly the Western economies. This is notwithstanding Azerbaijan's small oil resources compared to those of the Persian Gulf, which contains about 60% of the world's proven oil deposits and has phenomenal gas reservoirs (accounting for about 40% of the global deposits). Iran and Qatar, respectively, are the second-largest (29.61 trillion cubic metres (Tcm))[9] and third-largest (25.46 Tcm)[10] possessors of gas, with Russia (43.4 Tcm)[11] ranking first.

Consequently, the interest of the regional powers (Iran and Russia) and the non-regional ones (the US and EU), added to those of the Central Asian energy exporters (chiefly Kazakhstan and Turkmenistan), in the Caucasus lies in its potential as a land link between Asia and Europe and thus as an export route for the oil and gas resources of Central Asia, Iran and Russia. Needless to say, these countries have other interests such as stakes in developing and exporting its energy resources (the US and EU) and supplying fuel to Georgia (Iran) and Armenia (Russia and Iran).

For Iran and Russia to access international oil markets via their territorial waters, using the Caucasus for fossil energy exports is not vital. However, the situation is different for Central Asia and the United States, which have to rely on Iran and/or Russia for such exports. Accessing the Caucasus via the Caspian Sea will provide them with another export route. This is an especially important prospect for the Americans who wish to bypass Iran and Russia for exporting Central Asian and also Azeri oil and gas.

Compared to the Caucasus, Central Asia, which is prone to inter- and intra-state conflicts like the former, has much larger oil and gas deposits. Its two major fossil energy exporters, Kazakhstan (mainly oil) and Turkmenistan (mainly gas), have large energy resources. Their proven reserves in 2010 are respectively estimated at 39.8 billion barrels of oil[12]

---

[9] BP, "Gas: Proved Reserves", *BP Statistical Review of World Energy* (London: BP, June 2009), p. 22.

[10] Ibid.

[11] Ibid.

[12] BP, "Oil: Proved Reserves", p. 6.

and 1.82 Tcm of gas[13] and 0.6 billion barrels of oil[14] and 7.94 Tcm of gas.[15] These resources have enabled them to be players in the international markets for over three decades or so. Another regional state, Uzbekistan, possesses significant proven oil and gas reserves, i.e. 0.6 billion barrels of oil[16] and about 1.58 Tcm of gas,[17] enough to make it a major regional gas exporter as its oil production barely meets its domestic consumption. Energy resources are practically Central Asia's engine of growth and the main reason for the interest in that region of non-regional states, mainly the United States and the EU countries, although the region's border with two nuclear states (China and Russia) and a regional power (Iran) provides additional strategic interest.

Since their rise as oil and gas exporters in the post-Soviet era, the three Central Asian countries, which lack access to open seas, have relied on neighbouring Russia for their major exports through its extensive pipeline network inherited from the Soviet era. Yet, their almost exclusive reliance on Moscow for their vital energy exports have put the Russians in a position to exercise influence and impose any transit fee as they wish on the Central Asian exporters. Concerned about losing their real, but not nominal, independence, they have sought since the mid-1990s to find additional and/or alternative export routes to decrease their dependency on Russia. The recent instances of the latter cutting its gas exports via pipeline to Ukraine and Georgia in January 2006,[18] its piped oil exports to Belarus in January 2007[19] and again to Ukraine in January 2009[20] for a short while clearly indicated one thing. The Russians could use their oil and gas export infrastructure

[13] BP, "Gas: Proved Reserves", p. 22.

[14] BP, "Oil: Proved Reserves", p. 6.

[15] BP, "Gas: Proved Reserves", p. 22.

[16] BP, "Oil: Proved Reserves", p. 6.

[17] BP, "Gas: Proved Reserves", p. 22.

[18] "Russia Blamed for Gas 'Sabotage'",BBC News, 22 January 2006, http://news.bbc.co.uk/2/hi/europe/4637034.stm [Accessed 19 January 2010].

[19] "Russian Flows Via Belarus Halted", *Upstream*, 8 January 2007, http://www.upstreamonline.com/live/europe/article125800.ece [Accessed 9 September 2008].

[20] "The Gas Wars", *Newsweek*, 26 January 2009, http://www.newsweek.com/id/181720 [Accessed 1 February 2010].

through which other countries also export (e.g. Azerbaijan, Kazakhstan and Turkmenistan) or import (Georgia and Ukraine) energy as a means to exert pressure on those dependent on such networks. The completion of the first phase of the Central Asia–China Gas Pipeline in December 2009 connecting Turkmenistan to China via Uzbekistan and Kazakhstan, through which the three countries can export gas to China, ended Russia's monopoly on the Central Asian large gas exports, and added to the two Turkmen-Iranian short gas pipelines.[21] Yet, the regional gas exporters still conduct the bulk of their international exports via Russia.

Neighbouring Central Asia with its long border with Turkmenistan, Iran is the logical export route for the Central Asians as Russia is for its geographical location. It offers the shortest, the securest and the least expensive route to the international markets through its extensive oil export infrastructure. However, for obvious political reasons, Washington's opposition to Iran as a major or *the* major export route has forced the Central Asians to limit their oil exports through that country to swap deals of about 80,000 bpd in 2007,[22] the most recent year on which reliable statistics are available. Accordingly, Central Asia supplies Iran with their crude oil at Iran's Caspian Sea port of Neka, to be used by the Iranian oil refineries in the north, in return for the equivalent amount of Iranian crude oil delivered to designated customers at one of the Iranian Persian Gulf oil terminals.

Given that Iran is practically out of the question for Central Asia, the Caucasus's geographical location qualifies it as a potential transit route for not only Azerbaijan, but also for Central Asia. This is notwithstanding its being much longer, more expensive and less safe than the Iranian route. Since the required infrastructure could also be used for oil exports from Central Asia whose proven resources are much larger than those of Azerbaijan, the possibility of oil exports from Azerbaijan through Georgia (having access to the Black Sea) and its neighbouring Turkey (with access

[21] "China Opens New Central Asia Gas Pipeline", Reuters, 14 December 2009, http://uk.reuters.com/article/idUKLDE5BD02E20091214 [Accessed 1 February 2010].

[22] Energy Information Administration, *Country Analysis Brief: Caspian Sea*, January 2007, http://www.eia.doe.gov/cabs/Caspian/Full.html [Accessed 12 January 2010].

to the Mediterranean Sea) makes such exports even more interesting for the Azeris and the Georgians.

However, the wisdom of exporting oil and gas via the Caucasus, passing through Georgia as the region's connection with Azerbaijan, Turkey and the Black Sea, has been questioned in the aftermath of the August 2008 Georgian-Russian war.

Owing to the depth of hostility between Georgia and Russia and the unsettled issue of Georgia's two breakaway republics backed by Russia, Georgia could well be dragged into another war in which all pipelines passing through its territory could be damaged or destroyed with the effect of ending oil/gas exports for an unpredictable period of time. In fact, while BP denied any damage to its pipelines passing through Georgia during the August conflict, it accepted the closure of its major pipelines for unspecified reasons.[23] This included the closure of the Western Route Export Pipeline (WREP), which takes crude oil from Baku to the Georgian Black Sea port of Supsa, and the suspension of gas being pumped from Azerbaijan's Shah Deniz field into the South Caucasus Pipeline connecting Azerbaijan to Turkey via Tbilisi. BP attributed the closure of its main oil pipeline, the Baku-Tbilisi-Ceyhan, linking Azerbaijan to Turkey via Georgia, to "a fire".

With the strong possibility of intra- and inter-state conflicts, the Caucasus was not a desired route for energy exports even before the Georgian-Russian war. Today the argument in favour of this route is even less convincing in the wake of that war. Nevertheless, since the US opposition to the Iranian route is especially strong today, the Caucasus is the only remaining option for Central Asia to reduce its heavy dependency on Russia. Yet, large and long-term exports in this case require building a very expensive and difficult-to-maintain undersea oil pipeline to connect Central Asia to the Caucasus via the Caspian Sea. Any land pipeline would have to pass through either Russia or Iran which neighbour the two regions in the north and in the south, respectively. While an onshore pipeline is out of the question for the aforementioned reasons, there is no serious plan for a predictably very expensive offshore pipeline, a very

[23] Daniel Fineren, "Conflict Stems Georgia Oil, Gas Pipeline Flows", Reuters, 12 August 2008, http://uk.reuters.com/article/gc07/idUKLC53581720080812 [Accessed 14 January 2010].

controversial project due to its apparent environmental damage to the already highly damaged environment of the Caspian Sea. To this, one should add the fierce opposition of Iran and Russia to such a pipeline, not just for its anticipated environmental hazards, but also because it would undermine their national interests, depriving them of oil/gas exports through their countries.

Operating in this reality, Central Asian oil exports via the Caucasus are currently limited to those possible by using small oil tankers operating between the two regions across the Caspian Sea. However, the United States, whose oil companies dominate the Caspian oil industry, has been trying to convince the Central Asians to increase their exports via the Caucasus as part of its objective of bypassing Iran and Russia for Caspian oil and gas exports.

6.3 THE SALIENCE OF THE MAJOR CAUCASIAN PIPELINE PROJECTS

6.3.1 Azeri Pipelines

For the purpose of turning the Caucasus into a major transit route, certain pipelines have been constructed by the mainly Western oil/gas corporations. Being parallel pipelines constructed in the same corridor, the Baku-Tbilisi-Ceyhan oil pipeline and the Baku-Tbilisi-Erzurum (BTE) gas pipeline, also known as the South Caucasus Pipeline and the Shah Deniz Pipeline, are especially noteworthy because of their political and economic significance.

BTC pipeline: After years of negotiations which started in the 1990s, the 1,768 km pipeline project was completed on 25 May 2005.[24] Connecting Azerbaijan's oil fields to Turkey's Mediterranean port of Ceyhan via Georgia by land, the $3.7 billion BTC pipeline's construction was controversial given its heavy cost with no strong economic justification, its insecurity thanks to its passage through two countries (Georgia and Turkey) with serious sources of instability, including armed conflicts,

---

[24] Hydrocarbons Technology, *Baku-Tbilisi-Ceyhan (BTC) Caspian Pipeline*, http://www.hydrocarbons-technology.com/projects/bp/ [Accessed 14 January 2010].

and its extensive social and environmental damage on the affected countries. With an initial capacity of 1 million bpd, the pipeline went online on 13 July 2006. As stated earlier, its capacity is set to increase to 1.6 million bpd by the end of 2010.

Apart from its importance as an export pipeline, the BTC pipeline's significance lies in its political dimension. Accordingly, it is a blatant manifestation of the policy of bypassing Iran and Russia for Caspian oil exports. In particular, the US-backed pipeline project is meant to eliminate Iran as a logical export route for neighbouring Azerbaijan and to weaken its presence in Azerbaijan's oil industry as a means, among others, to eventually exclude Iran from the strategically important Caucasus. The American pressure on the Central Asian oil exporters to use the BTC pipeline for their oil exports also aims at denying Iran both economic and political gains in Central Asia. No wonder that the BTC pipeline is a potential source of dispute for Iran in its ties with Azerbaijan, Georgia and Turkey, through which the pipeline passes. The BTC pipeline is currently the most important one in the Caucasus for these reasons.

BTE pipeline: This is a major gas pipeline connecting Azerbaijan to Turkey via Georgia (442 km in Azerbaijan and 248 km in Georgia). The BTE pipeline links Azerbaijan's Caspian gas field of Shah Deniz to Turkey's gas pipeline network near its border with Georgia to initially supply customers in Azerbaijan, Georgia and Turkey.[25] Given the limited capacity of Turkey for additional imported gas, it is expected to eventually function mainly as an export route for Azeri gas. The $1 billion pipeline whose construction began in the fourth quarter of 2004 went online in July 2007.[26] The BTE pipeline is capable of carrying up to 7 Bcm of gas each year, with the possibility of doubling its capacity in the future. However, many technical problems have so far resulted in its numerous closures.

Other noteworthy Azeri pipelines passing through the Caucasus include a few shorter pipelines. Many of them are primarily meant to facilitate Azerbaijan's exports, such as the Baku-Supsa and Baku-Novorossiysk

---

[25] Kildrummy, *The South Caucasus Pipeline (SCP)*, 2006, http://www.kildrummy.com/about/casestudy.asp?csid=2 [Accessed 14 January 2010].

[26] Hooman Peimani, "A Global Update", *World Pipelines*, vol. 7, no. 12 (December 2007), p. 14.

pipelines connecting Azerbaijan's oil fields to the Georgian and Russian Black Sea ports, respectively.

6.3.2 Armenian Pipelines

Unlike Azerbaijan, Armenia and Georgia do not have any major oil and gas reserves to meet their own demand to a significant extent and export. Consequently, they have to rely on large oil and gas imports from the regional suppliers, Iran, Russia and Azerbaijan. Given this reality, the Caucasus has become attractive for these suppliers, especially Iran and Russia. They have practically started a quiet competition, oversupplying these Caucasian states, providing them with not only revenue, but also political influence. Hence, Russia's loss of market share enables Iran (having the world's second-largest oil and gas reserves) to fill the gap as a neighbouring state with friendly ties with both Georgia and Armenia.

Moreover, the geographical location of the Caucasus as a land link between Asia and Europe offers Iran an opportunity to use the region for energy exports to Europe via pipeline, a means to end its current reliance on Turkey for any such project in the future. As conflicts and wars in the Caucasus remove such opportunities altogether, Iran has a large stake in the region's stability. Among many other factors, including its friendly relations with Georgia, despite Georgia's pro-American orientation, and its concerns about Russia's resurrection, notwithstanding their current friendly relations, concerns about long-term instability in the Caucasus motivated Iran not to take sides with Russia in its August 2008 war with Georgia. Iranian Foreign Affairs Minister Manouchehr Mottaki's visit to Tbilisi following the war when Moscow was seeking Georgia's isolation manifested Iran's stake in the Caucasus's stability.[27]

Within this context, there are certain pipeline projects with the potential to develop into major ones. They include the Iran-Armenia pipeline for supplying Armenia with Iranian gas. The 142 km pipeline, of which

---

[27] "Mottaki: Tehran Closely Following up Caucasus Events", IRNA, 17 September 2008, http://www1.irna.ir/en/news/view/line-17/0809172789195647.htm [Accessed 18 January 2010].

42 km is laid in Armenia, was inaugurated in March 2007.[28] Yet, the actual exports started in May 2009.[29]

Through the pipeline connecting the Iranian gas export infrastructure near the Iranian city of Tabriz to the Iranian-Armenian border, Iran supplied its neighbour with up to 1.1 Bcm of natural gas in the first year. However, the pipeline is capable of handling up to twice as much in the future. Thus, it will enable Iran to export 36 Bcm of gas to Armenia over a 20-year period after the pipeline's second phase, which is currently underway, is completed. To that effect, Iran will increase the volume of its gas exports to 1.5 Bcm per year by 2011 to gradually be increased to 2.3 Bcm per year by 2019.[30]

The $220 million pipeline ends Armenia's heavy dependence on Russia for its gas imports, through a neighbour with friendly relations with the Armenians. This is especially important for Yerevan which is concerned about the tension-prone Georgian-Russian relations. In particular, in the post-Georgian-Russian war era, Georgia's hostile ties with Russia are predicted to last a long time, which makes it unwise for Armenia to depend on Russian gas supplies available via Georgia, which neighbours both Russia and Armenia. The gas provided through the Iran-Armenia pipeline is favourable for Armenia as Yerevan pays for it with electricity generated by the fifth unit of the Armenian Hrazdan power station, which was constructed and fully financed by Iran.[31]

In terms of its income-generating potential, the pipeline can be hooked up through a short link to Georgia and even be extended to Ukraine via Russia, a project in which both Tbilisi and Kiev have expressed interest to rid themselves of the heavy reliance on Russia for piped gas. Washington's

---

[28] "Iran-Armenia Gas Pipeline Inaugurated", IRNA, 19 March 2007, http://www2.irna.ir/en/news/view/line-18/0703190191144210.htm [Accessed 24 January 2010].

[29] "Iran Begins Gas Export to Armenia: NIGEC Managing Director", *Tehran Times*, 14 May 2009, http://www.tehrantimes.com/index_View.asp?code=194557 [Accessed 1 February 2010].

[30] "Iran-Armenia Pipeline Expected Online Soon", UPI, 19 May 2009, http://www.upi.com/Science_News/Resource-Wars/2009/05/19/Iran-Armenia-pipeline-expected-online-soon/UPI-40571242740949/ [Accessed 27 January 2009].

[31] Hooman Peimani, "The Iran-Armenia Pipeline: Finally Coming to Life", *Central Asia-Caucasus Analyst* (Washington, DC), 22 September 2004.

opposition aside, the Georgians and the Ukrainians must be very interested in the project now, what with the Georgian-Russian war of 2008 and deteriorating Russian-Ukrainian relations, which make Iran a very plausible energy supplier for both of them.

Finally, a major Iran-Armenian pipeline project, important for its political and energy dimensions, was concluded in December 2008. Accordingly, Tehran and Yerevan agreed to build a 300 km pipeline to transfer processed oil from Iran's Tabriz refinery to Araskh (Yeraskh) in Armenia.[32] Set to begin in the first half of 2010,[33] the US$200–240 million project will satisfy Armenia's annual requirements of 450,000–500,000 tonnes of fuel. Meeting all of Armenia's needs in imported processed fuels, the pipeline will practically end Russia's role as the supplier of such products along with its corresponding political influence in Armenia. The project once fully operational will certainly lift Iran's influence in neighbouring Armenia.

6.3.3 Georgian Pipelines

Georgia has been heavily dependent on Russia as its main supplier of fossil energy, particularly gas, since its independence in December 1991. The deteriorating relations between the two sides, especially since the Rose Revolution of 2003, have created serious doubts in Georgia about its continued dependency on Russia. Moscow's ending such exports in the aftermath of the Georgian-Russian war of August 2008 clearly revealed the depth of the problem.

It is no surprise that Tbilisi has been considering alternatives to Russia since 2004 in particular. As the construction of the Iranian-Armenian gas pipeline started in 2005, the development created a realistic ground for Tbilisi's optimism that its extension to that country could well become

---

[32] Hooman Peimani, "Middle Eastern Progress", *World Pipeline*, vol. 9, no. 5 (May 2009), p. 14.

[33] "Construction of Iran-Armenia Oil Pipeline to Start in 2010 — Minister", Arminfo, Yerevan, 20 November 2009, http://nl.newsbank.com/nl-search/we/Archives?p_product=NewsLibrary&p_multi=BBAB&d_place=BBAB&p_theme=newslibrary2&p_action=search&p_maxdocs=200&p_topdoc=1&p_text_direct-0=12C1A21C67FF3A30&p_field_direct-0=document_id&p_perpage=10&p_sort=YMD_date:D&s_trackval=GooglePM.

feasible thanks to its neighbouring Armenia and being on good terms with both Tehran and Yerevan. The latter was reflected in Georgian Prime Minister Zurab Noghaideli's reported discussion in March 2005 with his Armenian counterpart, Andranik Margarian, regarding connecting the Iranian-Armenian pipeline to Georgia.[34] Apart from Tbilisi's initial interest in diversifying its gas suppliers, the growing tension in Georgian-Russian relations prompted an expression of interest in the pipeline project.

The ascension to power of Georgian President Mikheil Saakashvili in January 2004 and his increasing the pro-American orientation of Georgia have contributed to deteriorating Georgian-Russian relations. This reality has made imprudent the current heavy reliance of Georgia on Russian piped gas, while making gas imports from Iran more attractive. It has been especially so since late January 2006 when the Russian-Georgian conflict over Russia's gas exports to Georgia unintentionally created a strong incentive for Georgia to view Iran, this time very seriously, as a necessary gas supplier, perhaps the one to replace Russia as the main supplier.

The Russian inability for about a week to supply gas to Georgia because of an "accident", claimed to be planned by the Georgians, forced them to sign a contract to immediately import gas from Iran via a pipeline (connecting Iran to Georgia via Azerbaijan) left idle for about three decades.[35] Iran started its exports on 29 January and Iranian gas (2 million cubic metres per day) entered the Georgian pipeline network the day after. Russia's resumption of gas exports on the same day did not change Georgia's plan of importing Iranian gas. While the unexpected expansion of the gas market was a sweet surprise for the Iranians, the real prize for Iran was the stronger-than-ever interest of Georgia to act on its long-discussed plan to import Iranian gas on a much larger scale by connecting its pipeline network to the Iranian-Armenian pipeline.

---

[34] Hooman Peimani, "Georgia and Ukraine: Buying Iranian Gas?", *Central Asia-Caucasus Analyst* (Baltimore), 6 April 2005, http://www.cacianalyst.org/view_article.php?articleid=3197 [Accessed 28 January 2010].

[35] "Caucasus: Georgia, Armenia Consider Options after Russia Pipeline Explosions", Radio Free Europe/Radio Liberty (RFE/RL), 1 February 2006, http://www.rferl.org/featuresarticle/2006/02/d2074170-d820-4948-812e-69551d17c950.html [Accessed 28 January 2010].

Obviously, Russia has used its gas exports to press for extracting concessions from the regional states with growing ties with Washington, as evidenced in its short-term announced closure of its gas pipeline to Ukraine in January 2006 over price disputes and its "forced closure" of its pipeline to Georgia because of an "accident". To this end, the closure of its pipelines to Belarus (2007) and to Ukraine (2009) should be added.

Unsurprisingly, Georgia has every incentive to end its heavy reliance on Russia for gas. Having had tension-free relations with Georgia since its independence in 1991, Iran is the best available regional gas supplier to help Georgia achieve that objective. This state of relations has continued despite the pro-American direction of the new Georgian government led by President Mikheil Saakashvili.

Iran and Georgia have therefore discussed further cooperation on energy, building on Iran's large exports to Georgia, on many occasions since 2006. Thanks to the existence of a Soviet-inherited gas pipeline network in Georgia, exporting Iranian gas to Georgia via Armenia would not require large investments as linking the Iranian-Armenian pipeline to that network would only involve a short connecting pipeline and a relatively small investment in other required infrastructure. Given the limited capacity of the Iranian-Georgian pipeline via Azerbaijan, Georgia's option of linking its gas pipeline network to the Iranian-Armenian pipeline seems too tempting to be resisted, especially when Tbilisi has no other serious option.

Given the circumstances, Washington's unhappiness with Tehran cannot be a very strong disincentive for the Georgians to prevent them from turning their pipeline dream into a reality. However, in the absence of a concrete agreement to date, the Georgians are yet to go beyond an expression of interest in the gas pipeline project, reach an agreement with Iran and Armenia, and ignore the expected American opposition to such agreement with its economic and political rewards for Washington's enemy, Tehran.

Finally, Russia's gas giant, Gazprom, opened a 150 km (Dzuarikau-Tskhinval) controversial gas pipeline in August 2009. It connects the Russian north Caucasus to South Ossetia, the breakaway region of

Georgia now recognised as an independent state by Russia.[36] This was notwithstanding Georgia's opposition to the pipeline, which further strengthens the region's dependency on Russia and consolidates the power and authority of the South Ossetia government while further weakening the breakaway republic's ties with Georgia. Through the pipeline, South Ossetia will annually import up to 252.5 million cubic metres of gas through the pipeline, which will end its dependence on Georgian gas supplies.[37]

6.4 BARRIERS TO THE CAUCASUS'S EMERGENCE AS A LONG-TERM EXPORT ROUTE

The Caucasus has the potential to benefit from oil as an exporter and/or a transit route. However, that possibility demands the prerequisite of durable peace and security to ensure the continued oil export from and through the Caucasus. In reality, all the regional countries have the seeds of inter- and intra-state instability, which could even spill over into neighbouring powers or which could drag them into armed conflicts in one form or another. Among various potential sources of instability, the following are prominent.

6.4.1 Potential for Armed Conflicts

Civil wars: The three Caucasian countries faced major armed conflicts upon independence. In Georgia, the rise of the two breakaway republics of South Ossetia and Abkhazia instigated bloody civil wars which ended in their practical, but not official, independence as ceasefire agreements terminated the conflicts in1992 and 1993, respectively. Armenia and Azerbaijan found themselves on hostile terms as a dispute over the independence from Azerbaijan of Armenian-dominated Nagorno-Karabakh

---

[36] "Gazprom Brings Onstream Dzuarikau — Tskhinval Gas Pipeline", *Gazprom News*, 26 August 2009, http://www.gazprom.com/press/news/2009/august/article67101/ [Accessed 27 January 2010].

[37] Ibid.

backed by Armenia escalated into a civil war in the post-independence era. Another ceasefire agreement ended the conflict in 1994, leaving 20% of the Azeri territory under the control of Nagorno-Karabakh's Armenian separatists. Owing to a prevalent sense of dissatisfaction with the status quo on both sides of the conflicts in all three countries, instability could well engulf the region. The fact that ceasefire agreements, instead of peace treaties to address their root causes, were used to stop the armed conflicts has been alarming. Such unsettled conflicts could and, if the current situation continues, will likely lead to another round of civil wars.

The unresolved territorial dispute between Armenia and Azerbaijan over Nagorno-Karabakh makes this scenario feasible. The frustration of both sides could lead to the resumption of civil war, pitting the Azeris against the Karabakhi Armenians to escalate the war between Armenia and Azerbaijan, both of which share borders with Iran. The Iranians could not be indifferent to such a war for many reasons, including their ethnic ties with both countries and the possible expansion of the war into Iranian territory adjacent to those countries. Moreover, this scenario, even in the absence of such expansion, would create serious challenges to Iran's border security, including an expected flow of arms into Iran. The inflow of war refugees would also be a source of concern for the Iranians, both for the financial costs and also the security implications of predictable weapon and drug smuggling into Iran from the belligerent countries as well as for its possible radicalising impact on the Iranians.

In the case of civil wars in the Caucasus, other regional powers (Russia and Turkey) and a non-regional power (the US) could well be dragged into the conflict in some form in support of one side or another. Given the growing dissatisfaction among the regional peoples with their ruling governments, intra- and inter-state wars could therefore engulf the entire Caucasus since a variety of unresolved ethnic and territorial issues have prepared the ground for such expansion, in addition to the existence of Azeris, Armenians and Georgians as minorities in their non-native regional countries. Iran could not remain impartial in a military conflict along its borders or somewhere in the proximity of its bordering region, a scenario which could drag Iran into a war with a strong possibility of resulting in a major regional war pitting Iran in some way against Turkey and the United States.

Conflict with Russia: Georgia and Russia have experienced unstable relations since the fall of the Soviet Union. They have become more unstable and hostile since the election of pro-American President Mikheil Saakashvili. The three major sources of tension in their bilateral relations have been Russia's role in Georgia's breakaway republics of Abkhazia and South Ossetia as their main backer, Tbilisi's demand for the closure of the Russian military bases in Georgia, and Russia's concern over the growing American military presence in Georgia. The Russian backing of those republics and the presence of the Russian forces there as peacekeepers have been the main obstacles facing the Georgian government in restoring its sovereignty over those regions.

In the post-Georgian-Russian war era, the Russian military bases in Georgia no longer constitute the main source of conflict in Georgian-Russian relations, since the last Russian base located in Georgia is already evacuated. However, the issue has turned into a far more important one for Georgia. Russia has since increased its military presence to 4,000 personnel in both Abkhazia and South Ossetia, a sharp increase from its former presence limited to a few hundred troops stationed in each of them as peacekeepers. Given the heavy presence of the Russian military in Abkhazia and its leadership's close ties to Russia, using Sukhumi's deep-water port as a base for the Russian Black Sea fleet is a distinct possibility for Russia, given its current unfriendly ties with and thus the uncertain status of its rental base at Sevastopol located in Ukraine's Crimea.[38] However, the situation will likely change for the better for Russia, considering the humiliating defeat of pro-American President Viktor Yushchenko in the 2009 presidential election. In the scheduled 7 February 2010 run-off presidential election, both contenders (Viktor Yanukovych and Yulia Volodymyrivna Tymoshenko) advocated ending hostility with Russia and expanding their bilateral relations.[39] Hence Russia may not need Sukhumi for its navy, at least in the foreseeable future.

---

[38] Author's note: This chapter was written three years ago when a pro-American/anti-Russian president was in power. At the time, Russian-Ukrainian ties were unfriendly and even hostile in some cases.

[39] "Ukrainian PM to Face Old Rival in Runoff Election", Reuters, 18 January 2010, http://www.reuters.com/article/idUSTRE60H3OU20100118 [Accessed 1 February 2010].

Being aware of his country's weaknesses vis-à-vis Russia, President Saakashvili has tried to use his government's growing ties, including those in the military field, with Washington to secure its pressure on Russia to end its backing of the breakaway republics and close its military bases. Keeping that idea in mind, in his address of 19 April 2005 to law students at Tbilisi State University, Georgian President Mikheil Saakashvili expressed his hope of ending Russia's influence in his country: "This year, for the first time in 200 years, we [the Georgians] can resolve the issue of pulling the Russian troops out of Georgia and Georgia's de-occupation once and for all".[40]

Two major factors deteriorating Moscow-Tbilisi relations since 2004 have been Russia's backing of Abkhazia and South Ossetia and Georgia's growing cooperation with the United States. In August 2008 the two countries fought a war, Russia recognised the declared independence of Abkhazia and South Ossetia, and Moscow and Tbilisi severed just about all their relations to become fully fledged enemies.

Given the prevailing animosity, Tbilisi's apparent intention to exploit a favourable international environment and particularly Washington's backing to force Moscow to withdraw its troops from Georgia's breakaway republics will likely lay the ground for a potentially dangerous escalation of another round of hostility between Russia and Georgia. Such a situation will have major implications for the stability of the Southern Caucasus and could well pull in, in one form or another, all the regional and non-regional powers with stakes in the Caucasus (Iran, Turkey, the US and EU).

6.4.2 Potential of Internal Instability

The threat of civil wars, while important, is not the only source of internal instability in the Caucasus. In particular, the growing popular discontent in Georgia and Azerbaijan is creating a suitable ground for social and political unrest. It is not a secret that the Azeri and Georgian governments

---

[40] Eurasianet, *RFE/RL: Georgian President Anticipates Decision on Closure of Russian Bases This Year*, 20 April 2005, http://www.eurasianet.org/resource/georgia/hypermail/200504/0037.shtml [Accessed 30 January 2010].

do not have a strong and reliable social basis. This reality opens the gates for the rise in various forms of civil unrest to destabilise the countries from within.

6.4.3 Potential Oil-Related Sources of Instability

The civil wars of the 1990s and the devastating Georgian-Russian war of August 2008 clearly indicated the conflict-prone nature of the Caucasus. The unresolved status of the disputes that prompted those conflicts makes the outbreak of new armed conflicts a distinct possibility in the foreseeable future. Against this background, energy-related issues could well add to the possibility of conflicts erupting in the region under certain conceivable scenarios. One is the continuation of a mainly Western, but not exclusively American, plan to exclude Iran and Russia from the region's energy industry, for instance by constructing pipelines through the Caucasus to bypass them for Caspian energy exports (e.g. the BTC pipeline). Added to many other sources of grievances in Tehran and Moscow in their relations with Washington, this could provoke them to make those pipelines unusable in some overt or, more likely, covert violent manner. That scenario could likely provoke a violent response to instigate a potentially major regional conflict. Another scenario is the conflict between Georgia and Russia over the latter's alleged intentional gas cuts to the former as experienced in January 2006. Of course, this requires as prerequisites the resumption of normal relations between the two countries, which were severed in August 2008, an unlikely scenario in the near future, as well as normal Russian gas exports to Georgia. As of February 2010 the fate of the Russian gas pipeline to Georgia and any Russian gas exports to that country is unknown.

Lastly, the absence of a legal regime in the Caspian Sea could well prompt conflicts. Since the 1990s, such absence has prepared the grounds for multiple ownership claims by the littoral countries on certain Caspian offshore energy fields, which developed into ownership disputes. As a clear example, the dispute between Azerbaijan and Turkmenistan has provoked a mini arms race between the two countries, while causing major diplomatic quarrels evident in them summoning their ambassadors from each other's capital. In addition, disputes between Iran and Azerbaijan over an oil field in 2001 had the clear potential of escalating into a military confrontation.[41] The Americans have since sought to beef up Baku's military. Added to the growing American military presence in the region and increasing hostility between Tehran and Washington, the persistence of ownership disputes has created a fragile situation conducive to armed conflicts backed by regional and non-regional powers.

6.5 CONCLUSION

The Caucasus has had more than its fair share of conflict and instability since the disintegration of the Soviet Union. The region could well witness a new round of intra- and inter-state armed conflicts potentially even more destructive than what it experienced in the early 1990s should the current course of events continue. Although this grim prediction could well become a reality, the Caucasus could also achieve a constructive and potentially prosperous future geared towards its oil and gas resources and also those of other Caspian states, namely Kazakhstan and Turkmenistan. Having brought a small degree of prosperity for Azerbaijan, the Azeri fossil resources will have a more tangible positive impact on the country as the Azeris increase their oil and gas exports. However, those resources are not large enough to secure Azerbaijan a long-term role as a major oil and gas exporter. Nonetheless, if invested wisely, their generated revenues will help, as a major factor, the Azeris address their various social, economic, military/security and environmental problems which were inherited from the Soviet era or which emerged in the post-independence era.

Moreover, oil-related projects could also significantly improve the lot of the entire region. Accordingly, large-scale Azeri and Central Asian oil and gas exports via the Caucasus, if they happen, will translate into various projects such as pipeline construction to generate income and employment for the entire region, directly and indirectly. Nevertheless, making this potential a reality will require addressing the looming threat of conflict and instability in the Caucasus. This is an absolute prerequisite as the highly realistic outlook of wars and civil wars in this region will

[41] Hooman Peimani, "Light at the End of Baku-Ashghabad Tunnel?", *Central Asia-Caucasus Analyst* (Baltimore), 16 July 2003, http://www.cacianalyst.org/view_article.php?articleid=1571 [Accessed 2 January 2010].

only discourage its use as a major transit route for Caspian oil and gas exports for all concerned parties, the Central Asian exporters and the major energy corporations operating in the region, alike.

As a result, unless the root causes of all regional disputes are addressed, the prospect of a regional oil boom prompted by major long-term Caspian oil and gas exports via the region will not be conceivable. Rather, war and instability will likely swallow up the whole region in the foreseeable future with the great possibility of dragging the regional and non-regional states that have a vested interest in the Caucasus into armed conflicts.

BIBLIOGRAPHY

BP. "Oil: Proved Reserves", in *BP Statistical Review of World Energy*. London: BP, June 2009.

BP. "Gas: Proved Reserves", in *BP Statistical Review of World Energy*. London: BP, June 2009.

"Caucasus: Georgia, Armenia Consider Options after Russia Pipeline Explosions", Radio Free Europe/Radio Liberty (RFE/RL), 1 February 2006,

http://www.rferl.org/featuresarticle/2006/02/d2074170-d820-4948-812e-69551d17c950.html [Accessed 28 January 2010].

"China Opens New Central Asia Gas Pipeline", Reuters, 14 December 2009,

http://uk.reuters.com/article/idUKLDE5BD02E20091214 [Accessed 1 February 2010].

"Construction of Iran-Armenia Oil Pipeline to Start in 2010 — Minister", Arminfo, Yerevan. 20 November 2009, http://nl.newsbank.com/nl-search/we/Archives?p_product=NewsLibrary&p_multi=BBAB&d_place=BBAB&p_theme=newslibrary2&p_action=search&p_maxdocs=200&p_topdoc=1&p_text_direct-0=12C1A21C67FF3A30&p_field_direct-0=document_id&p_perpage=10&p_sort=YMD_date:D&s_trackval=GooglePM.

Energy Information Administration. *Country Analysis Briefs: Azerbaijan*, October 2009, http://www.eia.doe.gov/emeu/cabs/Azerbaijan/Full.html [Accessed 27 January 2010].

Energy Information Administration. *Country Analysis Brief: Caspian Sea*, January 2007, http://www.eia.doe.gov/cabs/Caspian/Full.html [Accessed 12 January 2010].

Energy Information Administration. *Country Analysis Brief: Caspian Sea — Oil*, 2006, http://www.eia.doe.gov/emeu/cabs/Caspian/Oil.html [Accessed 10 December 2009].

Energy Information Administration. *Country Analysis Brief: Azerbaijan*, June 2005, http://www.eia.doe.gov/emeu/cabs/azerbjan.html [Accessed 10 December 2009].

Eurasianet. *RFE/RL: Georgian President Anticipates Decision on Closure of Russian Bases This Year*, 20 April 2005, http://www.eurasianet.org/resource/georgia/hypermail/200504/0037.shtml [Accessed on 30 January 2010].

Fineren, Daniel. "Conflict Stems Georgia Oil, Gas Pipeline Flows", Reuters, 12 August 2008, http://uk.reuters.com/article/gc07/idUKLC53581720080812 [Accessed 14 January 2010].

"Gazprom Brings Onstream Dzuarikau — Tskhinval Gas Pipeline", *Gazprom News*, 26 August 2009, http://www.gazprom.com/press/news/2009/august/article67101/ [Accessed 27 January 2010].

Hydrocarbons Technology. *Baku-Tbilisi-Ceyhan (BTC) Caspian Pipeline*, http://www.hydrocarbons-technology.com/projects/bp/ [Accessed 14 January 2010].

"Iran-Armenia Gas Pipeline Inaugurated", IRNA, 19 March 2007, http://www2.irna.ir/en/news/view/line-18/0703190191144210.htm [Accessed 24 January 2010].

"Iran-Armenia Pipeline Expected Online Soon", UPI, 19 May 2009 http://www.upi.com/Science_News/Resource-Wars/2009/05/19/Iran-Armenia-pipeline-expected-online-soon/UPI-40571242740949/ [Accessed 27 January 2009].

"Iran Begins Gas Export to Armenia: NIGEC Managing Director", *Tehran Times*, 14 May 2009, http://www.tehrantimes.com/index_View.asp?code=194557 [Accessed 1 February 2010].

Kildrummy. *The South Caucasus Pipeline (SCP)*, 2006, http://www.kildrummy.com/about/casestudy.asp?csid=2 [Accessed 14 January 2010].

Mossavar-Rahmani, Bijan. "The Challenge of the US Caspian Sea Oil Policy", *Motaellat-e Asyaie Markazi va Ghafghaz* [Central Asia and the Caucasus Review], 28 (Winter 2000), pp. 45–60. Quoted in Peimani, Hooman. *The Caspian Pipeline Dilemma: Political Games and Economic Losses*. West Port, CT: Praeger, 2001.

"Mottaki: Tehran Closely Following up Caucasus Events", IRNA, 17 September 2008, http://www1.irna.ir/en/news/view/line-17/0809172789195647.htm [Accessed 18 January 2010].

Peimani, Hooman. "A Global Update", *World Pipelines*, vol. 7, no. 12 (December 2007).

Peimani, Hooman. "Georgia and Ukraine: Buying Iranian Gas?", *Central Asia-Caucasus Analyst* (Baltimore), 6 April 2005, http://www.cacianalyst.org/view_article.php?articleid=3197 [Accessed 28 January 2010].

Peimani, Hooman. "Light at the End of Baku-Ashghabad Tunnel?", *Central Asia-Caucasus Analyst* (Baltimore), 16 July 2003, http://www.cacianalyst.org/view_article.php?articleid=1571 [Accessed 2 January 2010].

Peimani, Hooman. "Middle Eastern Progress", *World Pipeline*, vol. 9, no. 5 (May 2009).

Peimani, Hooman. "The Iran-Armenia Pipeline: Finally Coming to Life", *Central Asia-Caucasus Analyst* (Washington, DC), 22 September 2004.

"Russia Blamed for Gas 'Sabotage", BBC News, 22 January 2006, http://news.bbc.co.uk/2/hi/europe/4637034.stm [Accessed 19 January 2010].

"Russian Flows Via Belarus Halted", *Upstream*, 8 January 2007, http://www.upstreamonline.com/live/europe/article125800.ece [Accessed 9 September 2008].

"The Gas Wars", *Newsweek*, 26 January 2009, http://www.newsweek.com/id/181720 [Accessed 1 February 2010].

"Ukrainian PM to Face Old Rival in Runoff Election", Reuters, 18 January 2010, http://www.reuters.com/article/idUSTRE60H3OU20100118 [Accessed 1 February 2010].

CHAPTER 7

KAZAKH GAS POLICY IN THE CENTRAL ASIAN REGION: PROBLEMS AND PROSPECTS

Zhanibek Saurbek

7.1 INTRODUCTION

Following the demise of the Union of Soviet Socialist Republics (USSR) in 1991, 15 new independent states emerged. All relished the opportunity to operate as independent countries and to develop their economies in accordance with their own understanding of market structure. These new republics chose different paths towards the improvement and rehabilitation of the existing economies in their homelands. Naturally, after more than a decade, the resulting development levels of these economies are varied.

For the purpose of this article, an analysis of the relations between Kazakhstan and the main Central Asian republics in the energy sector will be provided. Moreover, other nuances such as political aspects will also be considered as politics and energy issues are closely interrelated, at least in the Central Asian region. Among the Central Asian republics which have been considered, we will discuss Kazakhstan, Uzbekistan

Figure 1 Commonwealth of Independent States: Central Asian States

Source: The University of Texas at Austin, University of Texas Libraries, Russia and the Former Soviet Republics Maps, http://www.lib.utexas.edu/maps/commonwealth/central_asian_common_2002.jpg.

and Turkmenistan. There are many reasons for cooperation between the Central Asian countries. Primarily, they are all linked geographically (Figure 1). Moreover, the previous Soviet economic and industrial structure of the market bound these nations within the USSR tightly together.

A good example, which characterised the Soviet economy, has been stated by energy experts:

> Political criteria such as integration of the political space of the Soviet Union played an important part (for instance when oil equipment industries were located in Azerbaijan to supply extractive industries in Western Siberia). Transport was essentially provided for free. Vocational decisions therefore could not reflect the costs of the transport, which — once priced

at cost and under constraints of a capitalist market economy — would have made many operations uneconomic.[1]

So, the economies of these republics were linked closely and the demise of the Soviet Union was painful for the newly emerged states as they were unfamiliar with the realities of the market.

Another reason for prospective cooperation is the cultural, ethnic and religious background common to the Central Asian nations. In addition, these countries have mineral resources and they were forced to cooperate with each other in one way or another in order to transport their energy sources to international consumer markets.

Among a number of energy sources, gas was considered as a case study. Why have relations between Kazakhstan and other Central Asian republics been examined with regard to the gas sector? Gas issues were chosen as a means of analysing the policy of Kazakhstan due to the fact that nowadays this source of energy is extremely politicised. The well-known Russian-Ukrainian gas conflict and so-called "gas wars" in 2006 and 2009 give a perfect example of how an energy source can be used as a political tool within mutual relations. Taking into consideration the fact that gas plays a significant role in Kazakh domestic consumption as well as in the international market, potential disputes and conflicts with Central Asian countries may arise in the very near future.

Analysis of these relations has been made through the tool of legal instruments: bilateral treaties, agreements and covenants. The effectiveness and implementation of these legal instruments help to demonstrate the genuine level of such cooperation and assist in understanding the intentions of both parties for future development.

7.2 KAZAKHSTAN

Kazakhstan is one of the fastest-growing economies in the Central Asian region among the post-Soviet republics. The republic possesses vast

[1] T. W. Wälde and C. von Hirschhausen, "Regulatory Reform in the Energy Industry of Post-Soviet Countries", *CEPMLP Journal* (1998), p. 5.

mineral resources, in particular oil, gas, coal, uranium and other sources.[2] At the same time, Kazakhstan faces the problem of how to deliver all those commodities to the international consuming markets, as the Kazakh network of gas pipelines is dependent on the Russian network.

Kazakhstan likes to regard itself as a regional leader among the other Central Asian republics due to a number of factors: large reserves of oil and gas in western Kazakhstan; soft investment climate; stability of the political, fiscal and legal regimes; and absence of ethnic conflicts and civil wars.

Kazakhstan chose a "multivectoral policy", which means it endeavours to shape good relations with the US, Europe, the Arab world and China at the same time as it enjoys excellent relations with Russia. In this context, Kazakhstan has developed some level of relationship with the main geopolitical powers of the world. It has concluded a number of agreements and treaties with different countries and became a member of various international organisations and initiatives. In other words, Kazakhstan is eager to be an active political and energy player and demonstrate to other neighbours in the Central Asian region its advanced position in the international community. As a result of such intensive participation in various organisations (in addition, Kazakhstan has been appointed Chairman of the Organization for Security and Co-operation in Europe (OSCE) in 2010), the conclusion of bilateral and multilateral agreements along with its investment climate and political and fiscal stability, a broad and extensive presence of different international energy players in Kazakhstan can be observed.

However, relations with its closest neighbours in the Central Asian region have a complex nature and demand a thorough analysis and knowledge of the real situation together with the historical background. An overview of the energy sector and analysis of the gas policy of Kazakhstan towards Uzbekistan and Turkmenistan helps us to see Kazakhstan's energy picture and to identify the genuine level of relations between these countries.

---

[2] For example, in 2008, the estimate of proved natural gas reserves in Kazakhstan was 100 trillion cubic feet (Tcf), which makes the country on a par with Turkmenistan — refer to the link http://www.eia.doe.gov/emeu/cabs/Kazakhstan/NaturalGas.html [Accessed 29 July 2009].

7.3 OVERVIEW OF THE GAS INDUSTRY OF KAZAKHSTAN

Kazakhstan has relatively large gas reserves in the western and southern parts of the country, with the most proven reserves being situated in the Karachaganak field. This oil and gas condensate field reportedly has proven natural gas reserves of 48 trillion cubic feet (Tcf). The consortium developing Karachaganak expects to produce 900 billion cubic feet (Bcf) by 2012.

Kazakh gas pipelines also lie mainly in the west. Generally, they are underdeveloped and indicate a lack of relevant infrastructure for domestic consumption:

> Through several pipelines originating in Turkmenistan and Uzbekistan passing through western Kazakhstan to Russia, the cities of Aktyubinsk, Uralsk, Kustanai and Rudny are supplied with Uzbek gas. These main transmission lines from Turkmenistan in particular are potentially available for access by producers in Kazakhstan to export gas to Russia or Europe. The majority of the gas shipped in these two main pipeline routes are transported into the Russian Gazprom system.[3]

The lack of gas supply infrastructure was justified by the fact that

> the pipelines were designed to meet the needs of industry and those of the residents of Kazakhstan were met only in the cities and towns along the transit pipelines. As a result Kazakhstan imports much of the gas it consumes.[4]

In addition:

> The trunk pipelines in Kazakhstan are not technologically linked, which prevents their use to pump inexpensive gas produced in the western regions of the country to the southern and northern regions of the country. This is especially a problem for consumers in southern regions and the city of Almaty.[5]

---

[3] K. Ostrander-Krug, "An Overview of Oil and Gas Pipelines in Kazakhstan", Denton Wilde Sapte report, 2002, p. 4.

[4] "Outlook for Kazakhstan's Gas Industry", *Alexander's Oil and Gas connections Journal*, 20 September 2006, http://www.gasandoil.com/goc/news/ntc54177.htm [Accessed April 2008].

[5] Ibid.

The domestic gas pipeline infrastructure serves only a few regions in the country, hence the majority of cities in the centre and north of Kazakhstan remain unconnected to the infrastructure.

As a result, those areas are vulnerable in terms of gas supply and serious problems were identified in domestic energy policy. Moreover, the south of Kazakhstan is a densely populated area, so the supply of gas to this region presents a significant challenge to the Kazakh government.

Another serious problem is the obsolete condition of the existing network of gas pipelines which needs to be modernised, replaced or reconstructed. In addition, new gas pipelines are of crucial importance for the government to diversify routes, develop infrastructure and connect to international markets. Fortunately, new pipeline projects to China (Turkmenistan-Kazakhstan-China) and to Russia (Turkmenistan-Kazakhstan-Russia) are being discussed and agreed between the parties and are likely to be realised in the medium term. The amount of gas transported is planned to be increased due to the soaring development of gas fields in the near future (Table 1).

Regarding domestic gas policy and supply, in its early days of independence, Kazakhstan had a bad experience with foreign management. After the breakup of the Soviet Union, the Kazakh economy was in a transition period with a number of different problems. One serious test for the newly emerged state was that of domestic gas consumption and gas supply. The gas pipeline infrastructure and supply literally were a burden to the state and it could not properly allocate funds and manage this asset as well as fulfilling its investment obligations and social responsibilities. In addition to those difficulties, consumers failed to pay promptly for their gas and the situation became disastrous.

Table 1 Volume of Gas Transported by Pipeline in Kazakhstan, 2002–2007

	2002	2003	2004	2005	2006	2007
Gas transportation by pipeline (Bcm)	110.5	12.1	121.6	13.2	121.4	124.8

Source: Presentation by K. Rakhmetova, Head of Transportation Dept., NOC KazMunaiGaz, on 8 November 2007 at a conference organised by ECT.

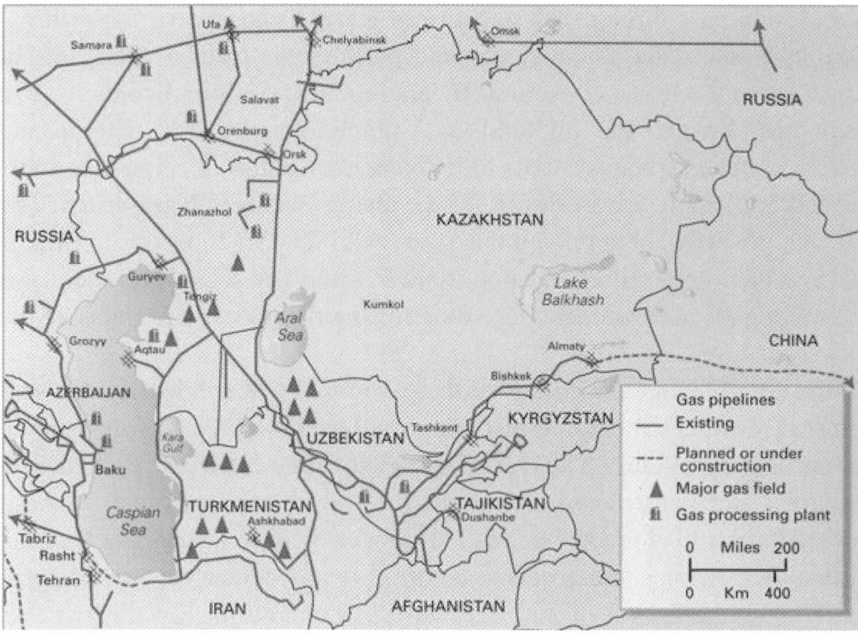

Figure 2 Natural Gas Infrastructure in Former Soviet Union (FSU) Central Asia

Source: Energy Information Administration (EIA) Statistics, *International Energy Outlook 2004: Natural Gas*, http://www.eia.doe.gov/oiaf/archive/ieo04/nat_gas.html [Accessed July 2009].

The Kazakh government decided to sell its gas trunk pipelines to a Western management operation (Belgian company Tractebel AC). This was in 1996–1997. Unfortunately, it was an unsuccessful experience because of the deterioration in the quality of the services they provided, drastic reduction in income for the state, decline in domestic gas consumption, and constant increase in tariffs for domestic consumers.

After long negotiations, the gas industry was taken back under state control and regulated through a specially formed gas company, KazTransGas (KTG). This company, together with other newly emerged companies, now controls the flow of gas through pipelines in Kazakhstan.[6] As can be seen from Figure 2, all the main gas pipeline routes from

[6]The official website of the company with historical background is available at http://www.kaztransgas.kz/article.php?article_id=667&mid=31 [Accessed April 2008].

Uzbekistan pass through the territory of Kazakhstan in a westerly direction, by passing the south, centre and north of the country. The west of Kazakhstan is a relatively self-sufficient region and is supplied by its own resources: Kazakh gas oil fields and pipeline routes from Russia and Asian republics (swap systems in both cases). So, the gas pipeline industry and gas supply are a priority for the energy sector of Kazakhstan. The Kazakh government approved the Strategy of the Gas Industry up to Year 2015, which emphasises the importance of the gas sector and identifies the main gaps and strategic decisions for the next decade to improve the industry.

To sum up with an overview of the gas industry, Kazakhstan possesses mineral resources including gas, but it suffers from a lack of gas supply to particular regions. The Kazakh gas pipeline network was mainly designed and constructed as part of the Soviet export system without consideration of domestic needs. Moreover, it depends on the Russian system of pipelines which acts as a barrier to any launch for international markets or the option of any alternatives. In addition, Kazakhstan suffered from the effects of its bad experience under foreign management for its domestic supply and so established a new national gas company. The legal background of the country within the gas sector is mainly oriented towards the satisfaction of domestic needs and reducing dependency on an external supplier. Among those external suppliers are Russia and Uzbekistan.

7.4 UZBEKISTAN

7.4.1 Historical Background

Uzbekistan is a close neighbour in terms of its common historical background, language, religion and geographical location. The two states have much in common as well as many differences. Initially, both Kazakhstan and Uzbekistan had similar conditions under which to develop their countries, but they chose different paths towards growth. Now their economic situations are quite different because Kazakhstan is more advanced with its economic reforms in such areas as foreign investment legislation, tax legislation, banking system reform and higher levels of investment and business. At the same time, "Uzbekistan's lack of progress in these reform areas has affected the level of investment in the country".[7] Uzbekistan has many peculiar features as part of its developmental and evolutionary process.[8] However, the two countries are interdependent in many areas, including the energy sector, so they are forced into a cooperative relationship.

In addition, Uzbekistan must cooperate with Kazakhstan because of the beneficial geographic location of Kazakhstan within the Central Asian region. Moreover, Uzbekistan has no physical links to Russia, which is a geopolitical power in the region as well as a former ally in the Soviet Union era.

7.4.2 Key Problematic Issues in Kazakh-Uzbek Relations

First, many problems related to regulations and political relations between the two republics can be identified. Kazakhstan depends on Uzbek gas in terms of the domestic gas supply to its southern region. Taking into account the inconsistency of the Uzbek position as a supplier, the process of gas supply to the southern part of Kazakhstan was, is and will be a very sensitive aspect of international relations over the next decade.

The south, centre and north of Kazakhstan are poorly served by the existing gas pipeline infrastructure. Moreover, there is no domestic pipeline from the western oil fields. Therefore, southern, central and northern Kazakhstan are dependent on external suppliers who are, as practice

---

[7] P. Blackmon, "Divergent Paths, Divergent Outcomes: Linking Differences in Economic Reform to Levels of US Foreign Direct Investment and Business in Kazakhstan and Uzbekistan", *Central Asian Survey*, vol. 26, no. 3 (September 2007), pp. 355–372.

[8] For an analysis of the characteristics of political elites within the country and its impact on the development of political institutions, see L. Adams, "Cultural Elites in Uzbekistan: Ideological Production and the State", in *The Transformation of Central Asia: States and Societies from Soviet Rule to Independence*, edited by Pauline Jones Luong (Ithaca: Cornell University Press, 2003), pp. 93–119; A. Ilkhamov, "The Limits of Centralization: Regional Challenges in Uzbekistan", in *The Transformation of Central Asia: States and Societies from Soviet Rule to Independence*, pp. 159–181; for complicated relations of Uzbekistan with the US and policy, see S. Akbarzadeh, *Uzbekistan and the United States: Authoritarianism, Islamism and Washington's New Security Agenda* (New York: Zed Books, 2005); for the place and role of Uzbekistan in the dimension of the international community, see M. Kirimli, "Uzbekistan in the New World Order", *Central Asian Survey*, vol. 16, no. 1 (1997), pp. 53–64.

shows, unpredictable, especially in the winter months and just as importantly in terms of tariffs.

The domestic consumption of gas is about 8 billion cubic metres (Bcm) per year. Consumption in the southeast and north of the country is catered for by imports from Uzbekistan (1.4 Bcm) and Russia (1.2 Bcm). Uzbekistan, in turn, is dependent on Kazakhstan due to the fact that the Central Asia–Centre pipeline runs through the territory of Kazakhstan. So, both countries are dependent on each other in one way or another. However, the dynamic of international relations does not show the positive changes in their mutual cooperation. Both leaders have their own views regarding the development of their relationship; however, it is difficult to say that those views entirely coincide. There is no doubt that in the Central Asia or post-Soviet political landscape much is dependent on the political and personal interaction between the leaders of states. History presents many cases where a political approach and good relationships can resolve issues without the additional mitigation of problems being required. Establishing mutually beneficial cooperation to the satisfaction of both parties is a very difficult task and as a result, many issues arise between the states in various areas.

Kazakhstan, within its associations with Uzbekistan, plays an unusual role: it plays the importer role rather than the exporter. This is quite unfamiliar to Kazakhstan, so it should learn from the lessons of international practice and identify the main priorities for developing its relationships with its closest neighbours.

No doubt, all arguable issues should be considered carefully and in a positive way. One of the ways to understand their mutual cooperation and to predict its future progress is to analyse the legal background between the parties. The level of achievement often can be observed from the agreements concluded between the states and the implementation of such agreements shows the intentions and motives of both parties.

7.4.3 Legal Instruments and Their Impact on Kazakh-Uzbek Relations

Due to the development of new oil and gas fields in Kazakhstan, improvements to and construction of new gas pipelines, providing gas to the south and centre of the country, as well as difficulties with existing contracts with Uzbekistan, the Kazakh government approved in 2002 domestic strategy principles called The Policy on the Development of the Gas Industry of the Republic until 2015 (decree no. 25) and The Program for the Development of the Gas Industry of the Republic of Kazakhstan until 2010.[9] Although gasification of the southern region was not stated there, many norms affect it indirectly, such as the development of the Amangeldy oil field (Zhambyl region) and the supply to the south of the country.

Paradoxically, a limited number of specific agreements were signed with regard to the gas supply to the south of Kazakhstan with Uzbek officials:

- The agreement between the government of Kazakhstan and the government of Uzbekistan on clearing off debts for fuel-energy resources for the years 1995–1996, and on the transit-and-swap system for the years 1996–1997.[10]
- The Protocol between the Customs Control Committee under the Ministry of State Income of Kazakhstan and the State Customs Committee of the Republic of Uzbekistan on customs control of natural gas, transferring through the customs border.[11]

Unfortunately, these and other documents failed to identify serious gaps between the two states on a political as well as a legal level.

In summary, with respect to gas, it should be noted that the south of Kazakhstan is and probably will continue to be dependent on foreign suppliers such as Uzbekistan and Russia due to a number of reasons such as undeveloped gas pipeline routes, the dispersal of gas to a third destination rather than for domestic consumption, etc. The situation with Uzbekistan will be complicated during the next three to five years. No new gas fields have been discovered and Kazakhstan is still vulnerable in that matter.

Another point to note is the possible routing of a gas pipeline to China from western Kazakhstan (with Turkmen and Uzbek involvement) and this could change the situation precipitately (Figure 3). The technical-economic

[9] Entered into by Decree of the Government No. 669, 18 June 2004.

[10] Made in Tashkent (Uzbekistan) on 18 November 1996.

[11] Signed in Tashkent on 26 November 2002.

Figure 3 Gas Infrastructure in Kazakhstan with the Proposed Route of the Kazakh-Chinese Gas Pipeline Project

Source: Ministry of Energy and Mineral Resources of the Republic of Kazakhstan, http://www.memr.gov.kz [Accessed June 2009].

feasibility of the project is considered possible by the construction of a branch of the trunk pipeline to Almaty (which is dependent from Uzbek gas) and some southern cities. In addition, a domestic pipeline will be built during the next three years to other southern cities to cover demand for natural gas. This project will be constructed using state investment.[12] Here Kazakhstan could use its advantage as a transit/supplier country to satisfy its own demand in gas. Kazakhstan could fulfil its own needs and also reap the profits from the transit and delivery of gas to China.

A further point is the potential for a liquefied natural gas (LNG) market which is, certainly, at this stage undeveloped. There are other risks, such as competing with existing players in the market and securing a relatively small share of the whole gas market. Perhaps it would be more profitable to extend this sort of business in the north and centre of Kazakhstan. Relations with Uzbekistan can be defined as a problem at a micro-economic level, because it mainly supplies gas for domestic consumption together with a small share through the Central Asia–Centre gas pipeline.

---

[12]This pipeline will be analysed later where the involvement of private participants, assumption of risk by the government and other issues are being considered.

Relations between Kazakhstan and Uzbekistan reflect their low level of mutual cooperation and weak legal background, because of complicated political relations, and as a result it influences the legal instruments and agreements. Kazakhstan, as well as Uzbekistan, is mainly oriented towards its own concepts and strategies to solve domestic problems. It is likely that the Asian gas pipeline project (Kazakhstan-China) will link all the parties involved in this project and may prove a connection for the further development of mutual relations.

The conclusion of various legal instruments and memoranda cannot promote economic activity or facilitate the easing of abrasive relations between the countries. That is why Kazakhstan's model of approaching domestic problems still centres on self-sufficiency and domestic-oriented policy rather than development of relations with Uzbekistan and the use of its energy resources.

The situation with Uzbekistan is of particular interest since its relationship with another Central Asian state, Turkmenistan, is quite different.

7.5 TURKMENISTAN

Turkmenistan is part of Central Asia and is an unusual country. Until recently, Turkmenistan chose a unique way to develop its economy and the welfare of its people, according to its own understanding of governance of the country.[13]

It was a closed country with its own prejudices, norms, regulations and morals for a long time after the Soviet Union's collapse. In addition, it is one of the countries which adopted political neutrality.[14] This country has

---

[13] Some recent discussions can be found in the following articles: A. Kuru, "Between the State and Cultural Zones: Nation-building in Turkmenistan", *Central Asian Survey*, vol. 21, no. 1 (2002), pp. 71–90; J. Anderson, "Authoritarian Political Development in Central Asia: The Case of Turkmenistan", *Central Asian Survey*, vol. 14, no. 4 (1995), pp. 509–528; International Crisis Group, "Cracks in the Marble: Turkmenistan's Failing Dictatorship", International Crisis Group Asia Report 44, 2003.

[14] Turkmenistan reaffirmed its neutral status on 12 December 1995 at the UN General Assembly. Please check the UN website or the official site of the former Turkmen President, http://niyazov.sitecity.ru [Accessed April 2008].

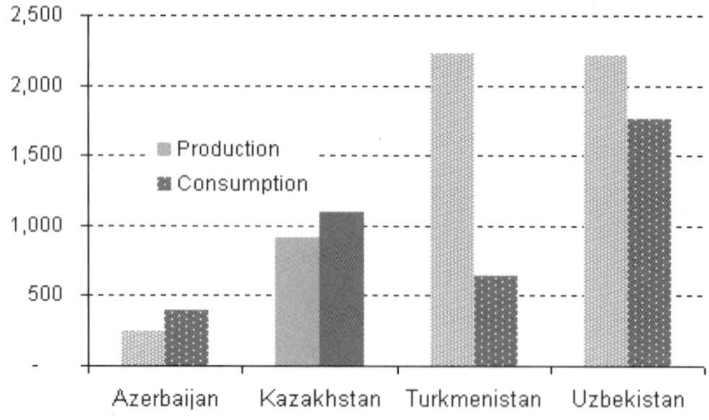

Figure 4 Central Asia Natural Gas Balance 2006 (Bcf)

Source: Energy Information Administration (EIA), *Natural Gas*, www.eia.doe.gov [Accessed July 2009]

huge gas reserves[15] (Figure 4) and also shares its borders with other Caspian littoral states.

The geographical location of Turkmenistan is significant because it offers a route into the international markets in a southerly direction. Kazakhstan has concluded several agreements with regard to oil and gas cooperation with Turkmenistan.[16] One of the first bilateral agreements was the Agreement on Collaboration in the Area of the Oil and Gas Industry between the Republic of Kazakhstan and Turkmenistan.[17] Then the presidents of both states agreed and signed A Joint Statement on Issues Associated with the Caspian Sea on 27 February 1997. Those agreements were mostly political-declarative in character and did not produce practical results.

Now Turkmenistan has changed dramatically after the demise of its first president, S. Niyazov, and the election of a new head of the country,

[15]Obtaining proper data on Turkmen gas reserves is highly difficult since state officials provide inaccurate figures; thus this is an unreliable source of information. But in accordance with BP, gas reserves are approximately 102,370 Tcf.

[16]For a full description of the legal regime in Turkmen's energy sector, see J. Hines and A. Marchenko, "Turkmenistan's Oil and Gas Sector: Overview of the Legal Regime for Foreign Investment", *JENRL*, vol. 24, no. 4 (November 2006), pp. 495–522.

[17]The agreement was signed in Ashgabat, on 19 May 1993 and came into force once signed.

G. Berdymukhammedov. Turkmenistan decided to open the country to positive relations with the international community and "stepped up a gear in its effort to deepen relations with the west in the wake of the death of the country's autocratic former president last December".[18] But Turkmenistan, at the same time, is trying to develop its relations with Russia and Kazakhstan.

As a result, Turkmenistan signed a number of bilateral agreements in different sectors with Kazakhstan[19] and became a strategic partner of Kazakhstan and Russia in respect of gas issues.[20] The recent signing of an agreement concerning gas transportation from Turkmenistan to Russia through the territory of Kazakhstan shows this new stage in relations between those three states.

Turkmenistan is a key producer of gas in Central Asia. Kazakhstan may be poised to act as a transit country to transport their resources in different directions, for example to China.[21] Thus, Kazakhstan needs to pay more attention to developing its relations with Turkmenistan and to transport its gas through their own territory. Turkmenistan, in its turn, has alternative options, such as Iran, for the transportation of its resources in different directions.[22] This would certainly be an undesirable option for the Kazakhs. In addition, the reverse might be a possibility where Turkmenistan

---

[18] "Turkmenistan Opens Gas and Oil Fields to the West", *Financial Times*, 27 September 2007.

[19] "Turkmen, Kazakh Presidents Sign a Number of Bilateral Agreements in Astana", 29 May 2007, http://www.turkmenistan.ru/?page_id=3&lang_id=en&elem_id=10145&type=event&sort=date_desc [Accessed April 2008].

[20] For a detailed analysis of Turkmen gas policy and the main challenges to this country, see M. Olcott, "International Gas Trade in Central Asia: Turkmenistan, Iran, Russia and Afghanistan", in *Natural Gas and Geopolitics: From 1970 to 2040*, edited by D. Victor, A. Jaffe and M. Hayes (Cambridge: Cambridge University Press, 2006).

[21] The Kazakhstan-China gas pipeline is also under construction; Turkmenistan and Uzbekistan are also involved into the project as additional suppliers. China has signed an agreement with Turkmenistan to produce gas there and transport it via a new pipeline that could open up the first non-Russian route for gas exports out of landlocked Central Asia. Please see "China in Gas Deal with Turkmenistan", *Financial Times*, 18 July 2007.

[22] For interesting opinions on the options of Turkmen gas in the context of the Turkey-Russia-Iran triangle, refer to R. Cutler, "Turkey and the Geopolitics of Turkmenistan's Natural Gas", *The Review of International Affairs*, vol. I, no. 2 (Winter 2001), pp. 20–33.

might act as a transit country with regard to the transport of Kazakh and Turkmen gas to Azerbaijan as an alternative route to Western markets.[23]

7.6 CONCLUSION

The international legal regime in Kazakhstan, as in other countries, cannot be operated in isolation of politico-economic, geopolitical, legal, cultural and other factors. All those components are integrated parts of the whole system of rules and principles, which reflect the will of the state.

Kazakhstan is pursuing its aim of becoming a regional leader through its reforms, by developing the level of its economy and democratic civil society. In addition, its mineral resources, international treaties, legal agreements and participation in different international initiatives, serious foreign investment and general approval of its current policy by the main geopolitical powers of the world, allow Kazakhstan to seek to be a significant independent energy player internationally, in general, and in Central Asia, in particular.

The Kazakh international legal framework includes various agreements and treaties with particular countries, international organisations and multilateral agreements. The main bias of the Kazakh energy transportation strategy is the creation of a multivector policy and the development of new alternative routes of transportation by pipeline or other methods.[24] Therefore, Kazakhstan is making significant steps towards the realisation of this concept. On the other hand, aiming to achieve such principles may lead to imminent conflicts of interest with different parties or geopolitical powers. However, it is worth noting that Kazakhstan is still dependent on

[23]For details of this specific project, see V. Socor, "Interest Rebounds in Trans-Caspian Pipeline for Turkmen Gas", *Eurasia Daily Monitor*, vol. 3, no. 16 (2006), http://www.jamestown.org/edm/article.php?article_id=2370697 [Accessed April 2008]; M. B. Chariyev, "Priority is Now Given to the Transcaspian Project", Caspian Energy, http://www.caspenergy.com/no4rus9.html [Accessed April 2008].

[24]The "Kazakhstan-2030" Strategy outlined a long-term way of development of the sovereign republic, directed at transforming the country into one of the safest, most stable, ecologically sustained states of the world with a dynamically developing economy. http://www.akorda.kz/en/category/gos_programmi_razvitiya.

its northern neighbour, Russia. Large investments from the US and the EU are another reason to create an effective model of relations. Kazakhstan also shares a border with rapidly growing China and its increasing demand for energy is a point difficult to ignore.

In fact, Kazakhstan has set an ambitious task — to keep a proper balance of interests in the region whilst retaining its own economic and politic dividends at the same time. For instance, Kazakhstan aims to be a key player in the system of transportation of hydrocarbons in Central Asia–Europe and Central Asia–China. But in negotiations with Ukraine, Kazakhstan has avoided committing to exact dates for the conclusion of any practical steps. As one expert correctly noted:

> Thus, Astana is always manoeuvring between Ukraine and Russia, aiming to consider its own interests in the energy sector. We may say that diversification of the market "in Kazakh" is an orientation towards Russia as the major transport corridor for oil and, simultaneously, launch in the markets of countries-consumers of oil and gas for the realisation of own projects. Hence, there are complicated combinations of issues for the Government of Kazakhstan.[25]

In addition to that statement, Kazakhstan is manoeuvring not only in that situation but in many others such as signing the Ankara declaration and joining the Baku-Tbilisi-Ceyhan system; the construction of the Kazakh-China oil pipeline; and giving consideration to the Trans-Caspian project. All alternative routes, as it can be seen, bypass Russia and reduce the dependency on its territory which in turn will reduce the influence of Russia on the energy policy of Kazakhstan.

With regard to the support of US interests in the region, it should be noted that the US is the largest investor in Kazakhstan. Astana signed the Ankara declaration and joined the Baku-Tbilisi-Ceyhan project. In addition, it allowed American companies to develop major oil fields in the western part of the country.[26]

---

[25] N. Kharitonova (2006). *Energeticheskii "treugolnik": Rossiya-Kazakhstan-Ukraina. (Energy "triangle": Russia-Kazakhstan-Ukraine)*, Lomonsov Moscow State University Faculty of History Journal, http://www.hist.msu.ru/Departments/CIS/Publ/2006_01_4.htm [Accessed April 2008].

[26] The Tengiz oil field was/is developed by the Chevron company.

Unfortunately, such an intensive dynamic cannot be seen in its relations with Uzbekistan. Turkmenistan is markedly different as views on and projects with Turkmenistan largely lie in the scope of mutual cooperation in alliance with Russia. Again, Turkmenistan has a significant geographical location and the ability to launch to the Iranian energy market. Iran plays a significant role in Kazakh energy transportation policy. Iran has a developed network of pipelines and a route to capacious international markets. Moreover, the uncertainty of a legal regime for the Caspian Sea where Iran is a littoral state, a new political regime and the prospect of an escalation of potential conflicts make this state of strategic importance. Iran could possibly be a strategic and tactical partner in the long term for Kazakhstan.

To sum up all those components, Kazakhstan with its geostrategic location in the Central Asian region could be a junction for the main transport routes through the Caspian states and one of the producers for the West and the East. That is why, at the moment, the most important thing is the development of a strategic partnership with Russia, China and other neighbours. Those countries and regions would have a high priority in the policy of transportation of hydrocarbons.

First, Kazakhstan should support the transportation of oil in multiple directions, develop an energy infrastructure, in general, and new routes and alternatives, in particular. Of course, it is a complicated and onerous task because it requires substantial financing and investment. But a regional and global cooperation with developed countries would partially solve this problem. As was mentioned, the "willingness of China, the US and Turkey (the EU position is yet to be seen) to pay for their 'strategic' interests by assuming the cost risks of pipeline infrastructure, cost overruns and external and internal security risk is well known. This would diminish the risk premium required for Kazakhstan projects".[27]

Second, one of the decisions may involve the usage and promotion of international legal instruments. For example, one international initiative which, without a doubt, will be useful and beneficial to Kazakhstan is the Energy Charter Treaty (ECT). The ECT has many positive effects and influences in the energy sector. As Kazakhstan is a country with an

[27] M. Ogutchu, "Kazakhstan's Expanding Cross-border Gas Links: Implications for Europe, Russia, China and other CIS Countries", *CEPMLP* (27 September 2006).

economy in transition, the level of political risk is still high and tangible. That is why the mechanisms of the ECT can regulate and mitigate potential disputes for foreign investors.[28] The role and effect of the ECT can be divided into two parts: for transit countries and for producing countries. Kazakhstan can be both. The ECT can give to Kazakhstan guarantees and benefits as a transit state.

In particular, it should be noted that different transport routes are being considered at present: Turkmenistan-Kazakhstan-China (gas pipeline) and Turkmenistan-Kazakhstan-Russia (also gas pipeline). Kazakhstan should coordinate and manage its mutual collaborations with regional states and partners more closely. The basis for further collaboration already exists. So the improvement of cooperation seems a strategic and significant goal. The progressive principles of the ECT were adopted and ratified by Kazakhstan, Uzbekistan and Turkmenistan. But instability in the political lives and regimes of neighbouring countries may reduce and delay further development of any projects, in particular, pipelines.

Third, Kazakhstan needs to improve its legal framework and develop legal mechanisms to avoid possible future disputes and even conflicts. All legal mechanisms such as participation in different organisations, initiatives and multi-bilateral agreements and treaties will only stabilise and improve any future collaboration. But the development of national legislation and modification in accordance with international norms and standards are an important and crucial aim.

Fourth, Kazakhstan has a strong position in terms of its location. None of the Central Asian republics have a physical border with Russia. By contrast, Kazakhstan has the longest border line with Russia in the world of 7,591 km. So, Kazakhstan should properly use this opportunity as a transit state. The agreements with Russia and Turkmenistan show the potential for Kazakhstan as a transit country. This role is a familiar one for Kazakhstan as it was a transit republic during the Soviet era and now must use this ability prudently.

---

[28]For detailed analysis of the investment process, refer to A. Seck, "Investing in the Former Soviet Union's Oil Industry: The Energy Charter Treaty and Its Implications for Mitigating Political Risks", in *The Energy Charter Treaty: An East-West Gateway for Investment and Trade*, edited by Thomas Walder (London: Kluwer Law International, 1996).

Lastly, experience with foreign management in the early days of independence forced the Kazakh government to open a new national gas operating company, which became a relatively successful entity in the country and now often operates as an independent player in international infrastructure projects. Perhaps it was a worthwhile lesson to understand that Kazakhstan should develop its own management with ideas learned from international practice.

To conclude, it is worth noting that Kazakhstan has a number of different problems in the gas sector, from difficulties in its relations with its closest neighbours and a weak legal background, to its dependency on a single country to provide gas for domestic consumption and the unsatisfactory technical conditions of its pipelines. However, Kazakhstan does have positive prospects in the gas sector from its excellent geographic location to future tangible gas projects. In that sense Kazakhstan should use those advantages in order to solve its domestic problems.

CHAPTER 8

THE THEORY OF STABLE ARAB GAS DIPLOMACY: REGIONAL ENERGY SECURITY THROUGH THE ARAB GAS PIPELINE

Mary E. Stonaker

8.1 INTRODUCTION

The role of infrastructure is striking in the emerging energy security cooperation between Middle Eastern nations. This correlation stems from factors that propel demand for geopolitical, economic and logistical integration on national and regional levels. Energy security encompasses traditional security on both levels as well. Through the establishment of physical cross-boundary infrastructure such as the Arab Gas Pipeline, energy security in the Middle East has been and will be encouraged.

There are currently two modes of transportation for natural gas; pipelines carry dry natural gas while ships and trucks carry liquefied natural gas (LNG). The stability of the pipelines as transit infrastructure serve better than ships or trucks to influence cooperation between nations despite tensions, conflicts or wars. Additionally, according to the International Energy Agency (IEA) and the Organisation for

Economic Co-operation and Development (OECD), "Long-term contracts [such as those signed when pipelines are choice of transit] have been a useful instrument to create a stable balance between gas exporters and importers".[1]

Though both modes of transportation are necessary to ensure diversified energy transport, there exist several reasons to favour pipelines as an accelerant to regional integration and stability. The security of pipelines can be controlled to a much greater extent than other vehicles of transit. In the event of an attack, the fallout from an "injured" pipeline will be much less than if the attack hit a truck or ship, especially when the latter vehicles pass through highly populated areas. Supply flow through a damaged pipeline may simply be turned off to prevent further leakage and repairs are often expedient.

This chapter will demonstrate that significance, defining a new theory called Stable Arab Gas Diplomacy (SAG Diplomacy), to investigate the role of pipelines in regional energy security as a powerful incentive. It will trace the history and current status of one such pipeline, the AGP.[2] Obstacles and risks faced by the countries involved in its genesis will be examined before evaluating its present status in the region. Market snapshots will summarise domestic natural gas markets in order to identify trends of increased natural gas consumption in relevant Middle Eastern nations. These increases unmistakably illustrate the growing incentive for regional cooperation. The prospects of the AGP will be looked at through the lens of current bilateral and multilateral discussions. Final remarks will collate the information presented into an analysis of the region's integrated energy security through the theory of Stable Arab Gas Diplomacy. The evaluation will conclude with a pivotal examination of inter-state tensions, conflicts and wars of such nations to provide a clear understanding of current regional energy security in the Middle East. It will be shown that, although the AGP will not provide a cure-all for Middle Eastern

---

[1] International Energy Agency, *Security of Gas Supply in Open Markets: LNG and Power at a Turning Point* (Paris: OECD/IEA, 2004), p. 23.

[2] The AGP was signed into existence in 2001, though the first operational leg was completed in 2003.

tensions, it will contribute to the alleviation of such tensions by encouraging regional integration.

8.2 THEORY

The various spheres of energy security produce many theories to explain and categorise national and international exchanges relevant to energy. There are three main strands from which these theories arise: international relations (IR), political science and economics. When narrowing focus regionally to the Middle East and topically to pipelines *as a source of diplomatic motivation* it becomes evident that, despite the wealth of theories existent in today's environment, there is not one that clearly applies to this hypothesis. According to one analyst, "Energy security analysis should be able to capture and combine both economic and political aspects of energy security as well as the perspectives of both energy producers and customers".[3]

Such an analysis is summarised succinctly as follows:

- **The theory of Stable Arab Gas Diplomacy**
  Assuming energy sectors are deregulated, transnational energy infrastructure is a striking incentive for Middle Eastern diplomatic integration. Such integrating infrastructure creates a stable and unified exporting market which produces positive externalities. Positive economic, political and social externalities from such an exporting market incubate stronger domestic economies. Over the long run, rather than turning inwards to compete over resources, the Middle East has been and will be drawn towards diplomatic cooperation to eliminate the economic, and perhaps eventually the political, sources of tension across the region.

It is important to note when evaluating the theory of Stable Arab Gas Diplomacy throughout this paper that it is a specialised look at one aspect

[3] Mikko Palonkorpi, "Energy Security and the Regional Security Complex Theory", University of Helsinki, 2007, http://busieco.samnet.sdu.dk/politics/nisa/papers/palonkorpi.pdf [Accessed 31 August 2010].

(transnational infrastructure) of one sub-sector (natural gas) of one sector (energy) of one region (the Middle East). The infancy in which the externalities from the AGP currently exist qualifies this as a preliminary look at best, the beginning of a generation-long investigation into the correlation between transnational infrastructure and regional integration. To sum it up, "The success of any integration scheme [...] enhances competition and efficiency within the integrated area, through increased specialization, and generally ensures better allocation of scarce resources into the most productive areas".[4]

8.3 ORIGINS OF THE ARAB GAS PIPELINE

While eyeing regional energy security integration, it is crucial that nations examine and enhance domestic security as well. Through sound energy security policies, which mandate deregulated energy sectors, the Middle East and its regional members will experience positive externalities. These include higher standards for services (for example, electricity), higher worker productivity and a stronger economy, which will lead to improved and increased economic and social infrastructure. Regionally deregulated sectors may also experience lower prices in the long term due to efficient market competition. On the other hand, if market competition is government regulated, countries will experience negative externalities due to such policies. Under regulation, prices are kept low as reserves and supplies are depleted at a faster rate, worker productivity decreases, infrastructure is left undeveloped or unused and such technical obstacles perpetuate the negative externality cycle. Without developing and maintaining a network of transnational infrastructure such as pipelines, regional integration and improved energy security is impossible.

There exists an argument amongst scholars as to whether external (extra-national) forces or internal (domestic) forces influence regional integration to a greater extent. The external forces argument holds that economies dependent on trade will visibly boost regional integration. Applying that argument to the Middle East, the construction

[4]Sunday Kachima McDonald Anadi, "Regional Integration in Africa", University of Zurich, April 2005, http://www.kfpe.ch/download/PhD_thesis_Anadi.pdf [Accessed 2 September 2010], pp. 26–27.

of transnational infrastructure and long-term bilateral and multilateral contracts will lead to greater regional integration. Alternatively, the internal forces argument cites a neo-classical theory: countries which adopt free-trade policies achieve high long-term growth which will spur regional integration. Working backwards, if nations aspire to regional integration, they should develop and adhere to similar frameworks in order to standardise bureaucracy and promote integration. For the purpose of this analysis, the first argument, the external forces argument, is more applicable.

The concept of a transnational natural gas pipeline originating in Egypt can be traced back to 1995. It was then that foreign oil companies were given permission to actively explore and drill for natural gas in the fields of Egypt in order to meet domestic consumption, standing at 439 billion cubic feet (Bcf) per year. In 1999, the Egyptian government declared that supplies of natural gas were being met. Egypt urged companies drilling for natural gas there to find an export market. The first export market for Egyptian natural gas was found in Jordan. After many logistical and cooperative discussions, it was decided through a memorandum of understanding (MoU) in 2001 that the Arab Gas Pipeline would be built to provide natural gas from Egypt to Jordan, Syria and Lebanon. Concurrently, an agreement with Israel found an offshoot of the AGP headed to Ashkelon, Israel, from Arish, Egypt. The AGP, which is 1,200 kilometres long and has a diameter of 36 inches, currently exports Egyptian natural gas and is divided into four segments. The first, from Arish to Aqaba in Egypt, was completed in 2003; the Aqaba-Amman-El Rehab section in Jordan in 2006; and the section from El Rehab, Jordan, to Damascus, Syria, and Homs, Syria, in February 2008. The Arish-Ashkelon extension of the AGP pipeline was also completed in the same month.

The AGP delivered its first 12 Bcf of Egyptian natural gas to Jordan in 2003. Since then Egypt's total exports of natural gas have risen nearly 500% to 598 Bcf in 2008 with the completion of further extensions to the pipeline along with completion of LNG infrastructure.[5]

---

[5] For more statistical data on Egypt's natural gas sector, please refer to Table 1.

8.4 OBSTACLES FACED BY THE ARAB GAS PIPELINE

There are several common barriers to investment and challenges faced during the genesis of the AGP project, including those that may be classified as market, regulatory, technical and political in nature.

Market risks faced by investors include the long payback period on the pipeline as an investment. The regional nature of this project poses risks because the investment is subject to the volatilities of many economies. There is no one international market for natural gas. As such, negotiations tackle the further chore of price-setting in addition to ownership/responsibility of gas while in transit. There are no international cartels or organisations governing the sale and purchase of natural gas either. A particular challenge to the AGP negotiations of 2000/2001 was the attacks of 11 September 2001. This event in the Western hemisphere changed the perceived economic landscape of the Middle East and dampened the investment spirits of American and European businesses in the region. Such events may be considered a political risk as well as a market risk.

Another market risk to investment is the degree of government regulation through mechanisms such as subsidies, which tend to be quite high in the Middle East.

In terms of regulatory risks, there exists no internationally binding legal framework governing the sale, transit and acquisition of natural gas in the Middle East. Although there are several frameworks currently being developed by the World Trade Organization and the Energy Charter Secretariat, none are binding as of publication date. Consequently, the nature and terms of natural gas agreements must be decided upon separately during each negotiation. Since 2000, negotiations have been undertaken in order to provide such a framework, known as the Transit Protocol,[6] by the Energy Charter Secretariat located in Brussels, Belgium. These negotiations have often been tabled due to disputes between the European Union and Russia.

---

[6]For more information on the Transit Protocol, please visit http://www.encharter.org/index.php?id=37

Technical issues that the AGP participants tackled include environmental challenges as well as the construction, maintenance and security of the physical pipelines and stations. These issues challenged not only the construction companies but remain an issue for all parties involved in the transit of natural gas through the AGP. A technical problem surfaced in 2008; Jordan was confronted with a sizeable shortage of natural gas when a pipeline malfunction disrupted supply. Jordan uses natural gas to run thermal power plants which generate electricity, the lifeblood of its economy. Due to this necessity for fuel to generate electricity, Jordan was forced to import heavy-oil fuel from Saudi Arabia in order to supply electricity grids of the country until the AGP once again became operational.

Political tensions of the region hindered negotiations during the creation of the AGP. These tensions will be examined later along with the AGP's positive effects on such tensions.

8.5 PRESENT STATUS OF THE ARAB GAS PIPELINE

Major shareholders in the Arab Gas Pipeline are EGAS[7] (Egyptian Natural Gas Holding Company) of Egypt, ENPPI[8] (Engineering for the Petroleum and Process Industries) of Egypt, PETROJET[9] (The Petroleum Projects and Technical Consultations Company) of Egypt, GASCO[10] (Egyptian Natural Gas Company) of Egypt and SPC[11] (Syrian Petroleum Company) of Syria. Other subsidiaries and companies with smaller shares come from the United States, Britain, Germany, Canada and Russia.

The volume of Egyptian natural gas flowing through the AGP is currently at 140 million cubic feet per year (MMcf/y) while its capacity is 350 MMcf/y.[12] In tandem with anticipating and urging the expansion of

[7] For more information, see Egyptian Natural Gas Holding Company (http://www.egas.eg).

[8] For more information, see Engineering for the Petroleum and Process Industries (http://www.enppi.com/).

[9] For more information, see The Petroleum Projects and Technical Consultations Company (http://www.petrojet.com.eg).

[10] For more information, see Egyptian Natural Gas Company (http://www.gasco.com.eg/).

[11] For more information, see Syrian Petroleum Company (http://www.spc-sy.com).

[12] Cristina Gallardo, "Shell Calls for Iraq's Gas Potential to be Unlocked", *Ordons News*, 5 April 2010, http://www.ordons.com/201004053857/shell-calls-for-iraqs-gas-potential-to-be-unlocked.html [Accessed 23 August 2010].

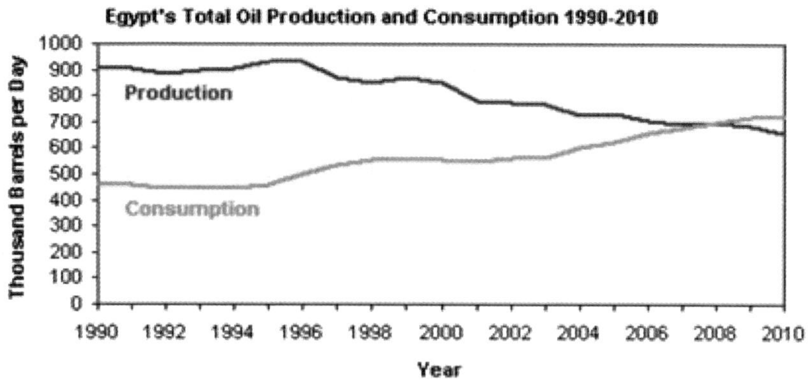

Figure 1 Egypt's Natural Gas Production and Consumption

Source: http://www.egyptoil-gas.com/read_article_issues.php?AID=477.

the AGP, connected countries will need to develop and/or streamline domestic infrastructure within a deregulated economic atmosphere in order to reap the greatest benefits from participation in the AGP project.

Figure 1 illustrates the surplus (exports) Egypt experienced, by year, in terms of its natural gas production and consumption. During 2008, Egypt was producing approximately 1.9 trillion cubic feet (Tcf) while consuming 1.1 Tcf.[13] Egypt possesses the third-highest estimated natural gas reserves at 58.5 Tcf in Africa, after Nigeria (185 Tcf) and Algeria (159 Tcf).[14] In 2009, Egypt exported 646 Bcf. Of the exports, 30% flowed out through pipelines while 70% was exported as LNG. Through the physical infrastructure of the AGP, Egypt is continually securing markets for export of its natural gas surplus in the Middle East as it eyes the markets of Europe.

8.5.1 Egypt

The Arab Gas Pipeline was born out of Egypt's Integrated Gas Strategy introduced by Sameh Fahmi who became Petroleum Minister in 1999.[15] It

[13]US Energy Information Agency: Independent Statistics and Analysis, *Natural Gas, Egypt*, June 2010, http://www.eia.gov/emeu/cabs/Egypt/NaturalGas.html [Accessed 20 August 2010].

[14]Ibid.

[15]"EGYPT — Integrated Gas Strategy; Exports Won't Exceed 25% Of Output", *APS Review Gas Market Trends*, 17 January 2000, www.allbusiness.com/mining/oil-gas-extraction-crude-petroleum-natural/409253-1.html [Accessed 3 September 2010].

was by the encouragement of Fahmi that President Mubarak diversified the energy sector through the construction of LNG ports and dry natural gas pipelines, including the transnational AGP.

Egypt discovered its large natural gas reserves in 1995 following permission granted to foreign companies to actively drill for hydrocarbons. In 1995, proven reserves stood at 25.20 Tcf. In August 2001, EGAS was incorporated as a state-owned entity. While many sectors of Egypt's economy have been deregulated since the 1990s, the energy sector remains under tight control with price ceilings and floors in place. Despite controls, foreign companies do have a presence in both the upstream and downstream sectors of natural gas in Egypt.

The AGP memorandum of understanding was signed in 2001 between Egypt, Jordan, Lebanon and Syria to provide an export market for Egypt's dry natural gas. The pipeline's construction was completed in different phases in 2003, 2006 and 2008.

Since 1994, proven Egyptian reserves have increased 400% from 19.2 Tcf to 77.20 Tcf in 2009; production of natural gas has risen 350% from 549 Bcf in 1995 to 1,935 Bcf in 2008; consumption of natural gas has risen 260% from 423 Bcf in 1995 to 1,108 Bcf in 2008; and exports of natural gas have risen from 0 Bcf in 1995 to 598 Bcf in 2008 (Table 1).[16]

Energy subsidies provided by the government defy the public sector and contribute to the instability of Egypt's energy sector. An example of such instability can be found in the blackouts (electricity failures) during the month of Ramadan in 2010. As a result of low prices, natural gas supplies were drained by the thermal power plants (generating electricity) and demand could not be met. Until Egypt allows prices to be regulated by the market, its domestic market will continue to face such shortages. Adjusting the volume set aside for domestic consumption may be necessary but before such measures are taken the market *must* be deregulated to allow for the most accurate assessment of domestic energy needs. Currently, thermal power plants in Egypt have switched from oil to natural gas and consume 65% of Egypt's total domestic natural gas supplies.[17]

---

[16] IEA country profile.

[17] MBendi, *Natural Gas Liquid Extraction*, http://www.mbendi.com/indy/oilg/gas_/af/eg/p0005.htm [Accessed 3 September 2010].

Table 1 Egypt, Natural Gas Volumes Per Fiscal Year

Year	Proven Reserves (TcF)	Produced (Bcf)	Consumed (Bcf)	Exported (Bcf)	Imported (Bcf)
2009	77.2	NA	NA	NA	NA
2008	75.95	1935	1108	598	0
2007	72.3	1872	1082	560	0
2006	68.2	1176	999	597	0
2005	66.3	1660	1208	293	0
2004	65	1296	1111	39	0
2003	59.4	1216	1046	12	0
2002	55.9	1107	883	0	0
2001	53.1	1083	867	0	0
2000	44.9	860	646	0	0
1999	36.4	694	518	0	0
1998	31	645	485	0	0
1997	27.6	641	477	0	0
1996	24.2	631	473	0	0
1995	25.2	563	439	0	0
1994	19.2	549	423	0	0

Table 1 is derived from information provided by the U.S. Energy Information Administration.
Note: Values set at zero are representative of the effects of proper infrastructure on exports (the AGP became operational in 2003) as well as the complete lack of imports.
Source: US Energy Information Administration.

8.6 SNAPSHOTS OF DOMESTIC NATURAL GAS SECTORS

Equally as important as Egypt's exports are to SAG Diplomacy are the domestic sectors of natural gas in the nations connected to the AGP. Despite the fact that there are forms of renewable energy being pursued in the Middle East, natural gas is the best candidate to replace oil and coal consumption, in the near future and on a large scale, due to its properties discussed at the beginning of the paper along with its cleaner burning properties with low carbon dioxide (CO_2) emissions. Especially relevant to Middle Eastern countries aiming to become net exporters, or increase net exports, of petroleum, shifting consumption to natural gas will allow these nations to export a greater amount of that more expensive fuel. The more oil available for export translates into more foreign currency injected into the national economies, an especially enticing prospect for these developing economies. Across sectors and borders, proper infrastructure and frameworks need to be created and upheld first in order to ensure sustainable growth. Only with such infrastructure will political tensions be eased, circumstances which will be analysed in a later section.

8.6.1 Jordan

Jordan lacks noteworthy petroleum and natural gas resources and therefore relies heavily on imports from its neighbours.[18] While continuing to search for investors to explore for greater reserves, Jordan transformed its electric power plants into natural gas-powered plants from diesel-powered in 2003. This swell in consumption can clearly be seen in Table 2; consumption jumped from the 2002 level of 11 Bcf to 23 Bcf in 2003, 55 Bcf in 2005 and 105 Bcf in 2008.

In 2001, the memorandum of understanding promised Jordan 1 Bcf/y of Egyptian gas (through the AGP which began operations in 2003); 96 Bcf worth of imports in 2008 supplemented the 9 Bcf produced by Jordan. Jordan has signed a US$3 billion 20-Year National Energy Development Plan which will deregulate its energy sector along with removing fuel subsidies.[19]

The added demand from Jordan's energy sector combined provides excellent incentive for integration and cooperation beyond its borders. A stable and prosperous Jordan will help ease tensions in the Middle East, especially the demarcation of borders. It will also ensure greater regional energy security and thereby ensure itself sufficient energy supply.

8.6.2 Syria

Syria, while pursuing potential investors to develop its own gas fields, has imported natural gas from Egypt through the AGP since 2008. Syria's

[18] Jordan has one developed gas field, near its border with Iraq.

[19] Encyclopedia of Earth, *Energy Profile of Eastern Mediterranean*, http://www.eoearth.org/article/Energy_profile_of_Eastern_Mediterranean [Accessed 25 August 2010].

Table 2 Jordan, Natural Gas Volumes per Fiscal Year

Year	Proven Reserves (Tcf)	Produced (Bcf)	Consumed (Bcf)	Exported (Bcf)	Imported (Bcf)
2009	0.213	NA	NA	NA	NA
2008	0.213	9	105	0	96
2007	0.213	9	92	0	83
2006	0.22	11	79	0	68
2005	0.22	10	55	0	45
2004	0.22	11	50	0	39
2003	0.23	11	23	0	12
2002	0.23	11	11	0	0
2001	0.23	10	10	0	0
2000	0.24	10	10	0	0
1999	0.245	10	10	0	0
1998	0.2	10	10	0	0
1997	0.2	10	10	0	0
1996	0.2	10	10	0	0
1995	0.2	10	10	0	0
1994	0.2	10	10	0	0

Table 2 is derived from information provided by the U.S. Energy Information Administration.
Note: The AGP became operational in 2003, an event marked by the sudden increase in imports. The table also explicitly outlines the absence of exports and trace amounts of proven reserves.
Source: US Energy Information Administration.

proven reserves have stalled at 9 Tcf since 1999 while its production and consumption have steadily increased (Table 3). Similar to its neighbours' policies, Syria has been attempting to shift domestic energy consumption from petroleum to natural gas to free up oil for export.[20] The AGP will serve to provide the necessary infrastructure to allow Syria's economy to develop beyond its own natural resources.

With political tensions stemming from the leadership of Syria, attracting investors has proved difficult though not impossible. In 2006,

[20] "The Energy Base of Syria", *APS Review Downstream Trends*, 8 March 2010, http://www.allbusiness.com/energy-utilities/oil-gas-industry-oil-processing/14068620-1.html [Accessed 20 August 2010].

Table 3 Syria, Natural Gas Volumes per Fiscal Year

Year	Proven Reserves (Tcf)	Produced (Bcf)	Consumed (Bcf)	Exported (Bcf)	Imported (Bcf)
2009	9	NA	NA	NA	NA
2008	9	208	213	0	5
2007	9	212	212	0	0
2006	9	221	221	0	0
2005	9	215	215	0	0
2004	9	251	251	0	0
2003	9	242	242	0	0
2002	9	240	240	0	0
2001	9	197	197	0	0
2000	9	215	215	0	0
1999	9	213	213	0	0
1998	8	203	203	0	0
1997	8	161	161	0	0
1996	7	142	142	0	0
1995	7	104	104	0	0
1994	7	134	134	0	0

Table 3 is derived from information provided by the U.S. Energy Information Administration.
Note: The AGP became operational in Syria in 2008, an event marked by the sudden increase in imports. The table also explicitly outlines the absence of exports. Despite high domestic consumption, Syria has exported small amounts of gas to Lebanon.
Source: US Energy Information Administration.

US-based Marathon Oil Corporation signed a gas (and oil) exploration deal with Syria. However, political tensions with the US and the speculation of sanctions caused Marathon Oil to sell its shares to Petro-Canada only months later in 2006. The estimates for the fields of Al Shae'r and Al Sharyfa stand at 80 MMcf/d with reserves of about 500 Bcf.[21] Petro-Canada has stated 2010 as the target on-stream date.[22] While exploration continues, the ability to tap into the permanent infrastructure of the AGP will provide Syria with a platform for stable development. Other factors

[21] Suncor, *International, Offshore*, http://www.suncor.com/en/about/919.aspx [Accessed 20 August 2010].
[22] Ibid.

would need to be considered in a future investigation to determine if Syria will effectively utilise such a platform for sustainable growth.

8.6.3 Lebanon

Lebanon was operationally linked to the AGP in 2006. Despite consuming and producing very little natural gas (Table 4), Lebanon is currently in the process of converting its power plants to run on natural gas from petroleum-based products. In 2006, Syria promised to supply Lebanon with 1.5 MMcf/d for 10 years. However, Syrian shortages of domestic supplies may hamper the fulfilment of that promise. Israel's offshore discoveries of

Table 4 Lebanon, Natural Gas Volumes per Fiscal Year

Year	Proven Reserves (Tcf)	Produced (Bcf)	Consumed (Bcf)	Exported (Bcf)	Imported (Bcf)
2009	0	NA	0	0	0
2008	0	0	0	0	0
2007	0	0	0	0	0
2006	0	0	0	0	0
2005	0	0	0	0	0
2004	0	0	0	0	0
2003	0	0	0	0	0
2002	0	0	0	0	0
2001	0	0	0	0	0
2000	0	0	0	0	0
1999	0	0	0	0	0
1998	0	0	0	0	0
1997	0	0	0	0	0
1996	0	0	0	0	0
1995	0	0	0	0	0
1994	0	0	0	0	0

Table 4 is derived from information provided by the U.S. Energy Information Administration (EIA).

Note: This table is provided, despite the utter lack of activity, to illustrate Lebanon's current standing in the region with regard to natural gas. Although consumption is slowly climbing, it is not climbing enough to register on the EIA's surveys.

Source: US Energy Information Administration.

natural gas fields may prove beneficial to Lebanon, if the maritime border dispute can be solved permanently.

Regionally, Lebanon holds a very critical geopolitical position, in large part due to the border it shares with Israel as well as the presence of Hezbollah. Hezbollah is backed by the Syrian government yet operates out of this small Mediterranean nation. Lebanon will play a key role in regional stability if reform and deregulation occur in its energy sectors. The creation of strong national energy security policies coupled with good governance reforms in Lebanon could help lead Middle Eastern regional integration in the right direction.

8.6.4 Israel

Rising Israeli demand for natural gas may be just what the region needs to provide a strong incentive towards regional integration for both Israel and its Arab neighbours. With steadily increasing production and consumption (Table 5), Israel may literally be sitting on a very powerful tool in terms of regional reconciliation. Of course, while it would be naïve and short-sighted to ignore the many other political issues to be settled between Israel and the Palestinian territories before full regional reconciliation could take place, integrating energy sectors would be a very strong step in the right direction. When approaching its newly discovered offshore gas fields, Israel must be careful to finalise maritime borders with Lebanon and Palestine before drilling. If it does not acknowledge its influential role in regional security by settling borders with Lebanon through the gas fields, Israel could very well further disrupt Middle Eastern regional stability beyond energy security.

Israel currently feeds about half of its domestic natural gas consumption, fuelling the other half through the Arish-Ashkelon pipeline. With a capacity of 335 MMcf/y, this pipeline stretches 100 km under the Mediterranean to connect Israel to the AGP. The pipeline was constructed and is operated by the East Mediterranean Gas Company (EMG). EMG is a multinational gas company and a joint effort of the Egyptian General Petroleum Corporation (EGPC) with 68.4% of shares, Merhav, an Israeli company, with 25% and the Ampal-American Israel Corporation with the remaining 6.6% of shares. The MoU signed between Egypt and Israel promised 63.45 MMcf/y; however, this has

Table 5 Israel, Natural Gas Volumes per Fiscal Year

Year	Proven Reserves (Tcf)	Produced (Bcf)	Consumed (Bcf)	Exported (Bcf)	Imported (Bcf)
2009	1.075	NA	NA	0	0
2008	1.075	99	42	0	0
2007	1.275	97	40	0	0
2006	1.375	35	34	0	0
2005	1.375	27	26	0	0
2004	1.375	28	28	0	0
2003	1.375	1	1	0	0
2002	1.47	(s)	(s)	0	0
2001	1.47	(s)	(s)	0	0
2000	0.01	(s)	(s)	0	0
1999	0.011	(s)	(s)	0	0
1998	0.011	1	1	0	0
1997	0.012	1	1	0	0
1996	0.013	1	1	0	0
1995	0.013	1	1	0	0
1994	0.014	1	1	0	0

Table 5 is derived from information provided by the U.S. Energy Information (EIA).
Note: Although consumption and imports are high, the EIA surveys do not reflect that. Such discrepancies may be a result of "regulatory risks" whereby the reporting of numbers is skewed or inaccurate because no legally binding framework exists within which to operate.
(s) — Numbers too small to represent in decimal format in the space allotted.
Source: US Energy Information and Administration.

since been increased to 78.65 MMcf/y through 2028.The original agreement sold the natural gas to the Israel Electric Corporation (IEC). In late 2009, EMG signed deals promising an additional 74.65 MMcf/y to private companies in Israel. IEC, on its corporate website, has predicted annual consumption will rise gradually until it reaches import levels of 149–223 MMcf annually.

8.6.5 Turkey

Turkey plays the role of infrastructural hub between Africa (Egypt), the Middle East, Europe and the Caspian Sea region. With the recently signed

MoU between Egypt and Turkey, the AGP will supply 100–400 MMcf/d to Turkey along with 203–608 MMcf/d to Eastern Europe, most likely through connecting to existing or developing pipelines such as the Nabucco Pipeline. The actual data is predicted to run at the low end of these promised volumes until recently discovered offshore fields come on-stream. Supplies may also be supplemented by other fields such as those in the Arabian Gulf and Iraq. This monumental agreement will no doubt be heralded by Egyptians but Turks as well. The Middle East's regional energy security will be strengthened significantly if Middle Eastern gas gains access to the European market. Such is the vital importance of developing permanent infrastructure to encourage long-term multi-regional energy security.

Turkey is motivated to participate in the AGP and Nabucco projects because it lacks any major proven reserves of natural gas (0.3 Tcf in 2009) and has very high consumption rates (1,238 Bcf in 2009) (Table 6). Turkey's economy will experience many positive externalities associated with pipeline construction and usage because it operates within a deregulated national energy sector.

8.7 EMERGING HORIZONS OF THE ARAB GAS PIPELINE

The future is bright — US$9 billion and 5 Tcf bright — for Egyptian natural gas, the Arab Gas Pipeline and regional energy security.[23] It was reported in July 2010 that BP and German RWE Dea will develop five offshore gas fields which will supply 1 Bcf/d with a target on-stream date of 2014. Rather than taking a percentage of total production as Egypt has done in past agreements with foreign companies, Egypt will pay two royalties for the gas; BP and RWE Dea will sell the gas to EGAS. The additional supply from these offshore fields will supplement other Egyptian natural gas production to boost supply to the domestic grid and the surplus will be exported through the AGP. Discoveries such as this coupled with the flexibility displayed by the Egyptian government will no doubt have a positive effect on regional energy security and integration.

[23]"More Gas for Egypt after BP, RWE Deal", *International Business Times*, 22 July 2010, http://www.ibtimes.com/articles/37592/20100722/more-gas-for-egypt-after-bp-rwe-deal.htm [Accessed 15 August 2010].

The AGP has also courted connections with other exporters and importers of natural gas. The exporting pipelines of Iraq (a deal signed in 2004 in tandem with Turkey) and the possibility of supplying the 3,300 km[24] European Nabucco Pipeline project[25] in conjunction with the future prospect of linking to the exporting Dolphin Gas[26] pipelines of Qatar and the United Arab Emirates will allow the region of the Middle East to target European natural gas markets as a stable and united net exporter. Working towards such levels of integrated regional cooperation signals that energy security may be the beginning of sustainable regional stability in the Middle East.

8.8 THE FIRST INDICATORS OF STABLE ARAB GAS DIPLOMACY

This section will collate the expansive set of information provided thus far with current snapshots of bilateral and multilateral interactions in the Middle East. This analysis of information will demonstrate the nascent correlation between the transnational infrastructure (AGP) and the long-term regional integration and energy security. While still in its infancy, the AGP has already shown signs that it is able to provide incentives for tension-bound neighbours to cooperate. Additionally, these nations need to view interdependence as a tool for regional growth rather than a national security threat. Aiming towards European markets, the AGP and the Middle East as a political, economic and cultural region will witness sustained growth *only* if it is interconnected and stable. The ability of the nations, including Israel, in this historically volatile area to collaborate on such a grand scale has already shown positive externalities.

---

[24] "Arab Gas Pipeline Approved by Turkey-EU", *Daily World EU News*, 6 May 2008, http://www.turks.us/article.php?story=20080506092427239 [Accessed 12 August 2010].

[25] "Minister: Egypt Considers Possibility of Transporting Its Gas on Nabucco through Pan-Arab Pipeline", *Trend*, 10 July 2010, http://en.trend.az/capital/pengineering/1718563.html [Accessed 14 August 2010].

[26] Dolphin Energy, *Dolphin Gas Project*, 2009, http://www.dolphinenergy.com/Public/facilities/facilities-gas-pipeline.htm [Accessed 14 August 2010].

The present status of regional affairs dictates further development of the AGP. In Egypt, it is an election year and a controversial one at that. Power cuts have plagued the country during the month of Ramadan and have prompted debates about the usefulness of the AGP and all natural gas exports. Water disputes between Syria and Jordan continue as cooperation on the AGP persists. Border rows between Jordan, Syria, Lebanon, Israel and Egypt linger in the background of energy talks. The sustained cooperation on joint energy security via the AGP clearly demonstrates the positive relationship between the stable infrastructure of the pipeline (and the natural gas its carries) with improved regional integration. Although there are still many tensions existing in other realms of the relationships of the region, the fact that all governments are cooperating on a project of this magnitude confirms progress in the right direction.

8.8.1 Close Neighbors Economic and Trade Partnership Council (CNETAC)[27]

On 10 June 2010, four of the five nations attached to the AGP (Turkey, Jordan, Syria and Lebanon) signed a free trade agreement, dubbed the Close Neighbors Economic and Trade Partnership Council (CNETAC). The CNETAC aims to "strengthen the existing cooperation, develop long-term strategic partnership and solidarity, and further improve the economic integration".[28] The first committee meeting will be held in September 2010 in Amman, Jordan, to sketch a roadmap of aims prior to the first ministerial meeting planned for Damascus, Syria, in December 2010.

This clear step towards regional integration supports the Stable Arab Gas Diplomacy theory as all countries are connected via transnational energy infrastructure. Turkish Foreign Trade Minister Zafer Çağlayan wrote in an official statement, With this close cooperation, our goal is to increase and diversify trade and investments among the four countries by

---

[27] "Turkey, Syria, Jordan, Lebanon on First Step to Set Up Mideast's 'EU'", *World Bulletin*, 1 August 2010, http://www.worldbulletin.net/news_detail.php?id=62093 [Accessed 29 August 2010].

[28] Ibid.

creating a liberal trade and investment environment with a *modern infrastructure at the international level*, free from all tariff and non-tariff barriers, encompassing a geography feeding a population of 105 million and, as of 2009, having a combined gross domestic product [GDP] of $723 billion, imports amounting to $176 billion and exports to $131 billion.[29] (Italics inserted by author.)

8.8.2 Regional Israeli Relations

Although Israel's neighbourhood relations have been stressed to say the least since its controversial creation, there are signs to indicate Israel is seeking détente. Pragmatically, this détente could be attributed to a series of shifts or events; however, it does coincide with the creation and evolution of the AGP as its consumption of natural gas well exceeds its domestic production. Spanish Foreign Minister Miguel Angel Moratinos acted as liaison between Israel and the states of Syria and Jordan through the media in May 2010. Following a tour of the Middle East, Moratinos was quoted as saying, "Israeli authorities asked me to convey a message to Syria and Lebanon, which is that they seek to ease tension, and they expressed their willingness for negotiations".[30] Israel began receiving imports through the AGP on 1 May 2010.[31]

In the spirit of unbiased evaluation, it is also critical to acknowledge the continuing existence of deep-set tensions in the region. The recent discovery of major gas fields offshore from Israel has stirred up tensions with its neighbour Lebanon and the militant group Hezbollah which resides in Lebanon. Executive Director of the Institute for the Analysis of Global Security, Gal Luft, recently wrote, "As a militarized non-state actor,

---

[29] "Major Step Taken towards M. Eastern Economic Union", *Today's Zaman*, 2 August 2010, http://www.todayszaman.com/tz-web/news-217874-105-major-step-taken-towards-meastern-economic-union.html [Accessed 29 August 2010].

[30] "Israel Seeking to Ease Tension with Lebanon, Syria: Spain", *EU Business*, 13 May 2010, http://www.eubusiness.com/news-eu/mideast-diplomacy.4p5 [Accessed 4 September 2010].

[31] Israel Electric Company, *Fuel*, http://www.iec.co.il/bin/ibp.jsp?ibpDispWhat=zone&ibpDisplay=view&ibpPage=IRRWPage&ibpDispWho=FuelIRR&ibpZone=FuelIRR& [Accessed 7 September 2010].

Hezbollah has already demonstrated its capability to push an independent agenda which serves its patrons Iran and Syria, not the people of Lebanon".[32] At the same time, he wrote, "Israel has already announced its willingness to use force to protect its natural gas finds".[33]

8.8.3 Syrian-Lebanese Relations

In early September of this year, Lebanese Prime Minister Saad al-Hariri retracted his charges against Syria for the death of his father, Rafik al-Hariri, in 2005. Outcries from local and international populations prompted Syria's President Bashar al-Assad to withdraw Syrian troops from Lebanon in the same year. This unthawing of relations between the two countries is encouraged by small amounts of Syrian natural gas exported to Lebanon through the AGP. (Lebanon also receives Egyptian gas through the AGP.) An Al-Jazeera correspondent in Beirut commented on the recent developments, saying, "Now Syria has emerged from its isolation as a very strong player in the region ... and so he [Hariri] has to adjust."[34] This shift in regional politics coincides with the on-stream status of the AGP since 2005, along with PM Hariri's election to office in that same year.

Without overstating the symbolic and pragmatic importance of the AGP in regional security, the recent warming of relations between Syria and Lebanon will continue to be encouraged by the physical infrastructure supplying countries the lifeblood of any economy: energy. The management of such resources both nationally and regionally is a topic for another day.

8.8.4 Syrian-Turkish Relations

Turkish-Syrian tensions centre around two issues: the Kurdistan Workers' Party (PKK) rebels and water rights. The southern neighbour showed the

---

[32] Gal Luft, "Hizballah Takes Aim at Israel's Natural Gas Discovery", *The Washington Times*, 13 July 2010.

[33] Ibid.

[34] "Lebanon PM Retracts Syria Charge", *Al Jazeera English*, 6 September 2010, http://english.aljazeera.net/news/middleeast/2010/2010/2010/2010/09/201096114429752335.html [Accessed 6 September 2010].

first signs of rapprochement in 1998 when Syria deported PKK leader Abdullah Ocalan. More recently, Syria's recent offer to give asylum to 1,500 PKK rebels if they renounced their armed conflict was another sign of improving relations.[35] In a 2009 meeting, the two nations addressed the water dispute with a "friendship dam agreement" although tensions still exist today over volumes reaching Syria.

8.9 CONCLUSION

While nations around the world are shifting from petroleum to natural gas consumption (at least in part), it is crucial that the Middle East integrate as a region to capitalise on this demand. Having long-term terrestrial access to the markets of Africa, Asia and Europe, the Arab Gas Pipeline is poised to be a great diplomatic tool for the Middle East to unite around a common interest. It is important to keep in mind that it is difficult, if not impossible, to divorce the economic features from the political aspects of energy security. Overall, the AGP has and will continue to produce positive externalities domestically and regionally for the Middle East and its member nations. The diplomatic incentive is the tool focused on in this paper through the definition of a new theory, the theory of Stable Arab Gas Diplomacy.

SAG Diplomacy is dependent on deregulated energy sectors, of which the Middle Eastern nations may not fully possess. However, it is still appropriate to begin tracking the effects of transnational infrastructure as these energy sectors and economies slowly move towards integration with the constant incentive of permanent infrastructure linking the nations. The AGP will provide the most incentive if and when nations adopt deregulated energy sectors. Nevertheless, it is evident that SAG Diplomacy is already at work in the region.

The best example of SAG Diplomacy, to date, is the formation of the CNETAC, a multilateral (regional) free trade agreement. Although also in its infancy, the CNETAC provides definitive support to the theory proposed in this paper. Other examples provided earlier in the paper

[35] "Turkey and Syria Signal Improved Relations", *New York Times*, 13 October 2009, http://www.nytimes.com/2009/10/14/world/europe/14turkey.html [Accessed 29 August 2010].

illustrated bilateral signs of the beginning of appeasement. An example of simple positive externalities of the construction and maintenance of the AGP is the creation of jobs which has injected spending power into the economy.

Secure energy infrastructure causes demand to increase, which in turn will create more jobs and a stronger economy. These stronger economies will be in a position to produce strong political bureaucracies which will serve to spur further investment into the region. Some risks discussed include the lack of binding regulatory bodies. Such frameworks, which have begun and continue to develop, will increase investor confidence in the region.

Through greater regional stability and integration, the Middle East will position itself to capitalise on access to foreign markets, especially in Europe. As Europeans fret over a lack of diversification in gas supply routes, the AGP now has perfect timing for Middle Eastern gas to "walk through that open door".[36]

While the diplomatic weight of regional energy security is heavy, it is not enough on its own to fully ease tensions of the region, in particular political tensions. This chapter has illustrated the platform created by the Arab Gas Pipeline, through the eyes of SAG Diplomacy, on which agreements such as the CNETAC have been made.

[36] For an analysis of European concerns, see Kevin Rosner, "Closing the Gap between Energy and National Security Policy", *Journal of Energy Security* (18 May 2010), http://www.ensec.org/index.php?option=com_content&view=article&id=245:closing-the-gap-between-energy-aamp-national-security-policy&catid=106:energysecuritycontent0510&Itemid=361 [Accessed 29 August 2010].

INDEX